D1766055

THE EXERGY METHOD
OF ENERGY SYSTEMS
ANALYSIS

The Exergy Method of Energy Systems Analysis

JOHN E. AHERN

Aerojet ElectroSystems Company

A Wiley-Interscience Publication

JOHN WILEY & SONS, New York · Chichester · Brisbane · Toronto

Library of Congress Cataloging in Publication Data:

Ahern, John E   1921–
    The exergy method of energy systems analysis.

    Bibliography: p.
    includes index.
        1.   Power (Mechanics)   2.   System analysis.
    3.   Thermodynamics.   I.   Title.

TJ163.9.A3   621.4   79-24500
ISBN 0-471-05494-1

Printed in the United States of America

10 9 8 7 6 5 4 3 2 1

To Cora *and*

       Tom

       Maureen

       Corinne

       Kathleen

       Tim

       Jeannine

Preface

The primary objective of this book is to provide engineers with a method of applying the second law of thermodynamics to their analysis and design of energy-related systems. The exergy method of analysis is relatively easy to apply to systems analysis, and the results will reward the engineer with greater knowledge of the real system characteristics. The history of the second law of thermodynamics is fascinating. The second law will not be extensively used, however, unless application demonstrates that it is worth the added analytical effort.

Energy conservation is a key goal of our economy now and will continue to be in the future. Aside from the drastic measure of slowing our economy by force, the most effective way to reduce energy demand is to use energy more efficiently. Many scientists now recognize that the second law of thermodynamics can help them discover where our available energy is inefficiently used. The waste of energy results in a partial loss of the work that is available from our limited energy resources. The second law of thermodynamics can realistically determine the magnitude and type of these losses, whereas the conventional heat balance of the first law cannot.

The second law of thermodynamics has been studied extensively since the days of Carnot, Clausius, and Kelvin. Thermodynamic textbooks include sections on the second law, usually based on a classical approach, which tries to prove the principles of reversibility and irreversibility of processes and cycles by complex and, in some cases, devious physical and mathematical proofs. Statistical and probabilistic techniques have been introduced to support proof of the second law of thermodynamics and to define entropy. These approaches have been analyzed by physicists specializing in thermodyamic concepts, but they have not been easily accepted by engineers who are ready to apply these principles to the design and performance analysis of their systems.

One approach that has been widely accepted in the Soviet Union and Europe is the exergy method of analysis. This method applies the second law of thermodynamics to energy-related systems, but does not require much of the complex thermodynamic analysis associated with reversibility and entropy. This book presents the exergy method of analysis from the standpoint of its application in engineering. The relationship of exergy to thermodynamic reversibility and to entropy is discussed only to the extent needed to apply the exergy method to actual energy-related systems. For further study of the second law of thermodynamics, the reader is referred to the bibliography.

Chapter 1 introduces the second law and exergy; a discussion of the background work performed in this field follows in Chapter 2. The physical basis of exergy is presented in Chapter 3, which, although important for the most effective use of the exergy method, is not necessary for preliminary exergy analyses. The procedure for applying the exergy method is given in Chapter 4, the building block scheme of processes allows any energy system to be analyzed by selecting and assembling the equation blocks to the particular system. In this way the analysis can be made to any degree of approximation or complexity. In Chapter 5 the procedures are applied to actual energy-related systems to demonstrate the use of the procedures. This chapter also compares the results of the exergy analysis to the conventional heat balance method to show the benefits of using the exergy method. Chapter 6 reviews potential applications of the exergy method relative to energy conservation and to improvement of system efficiency. The design of new systems using the exergy method as well as ways to improve existing systems are also discussed. Chapter 7 provides tables and charts of exergy values for many common working fluids in energy systems. The Appendix includes several translations of Russian papers that discuss the application of the exergy method.

I had initially planned simply to collect translations of Russian papers on exergy. In discussions with other engineers, however, I realized that only a thorough presentation of the procedures for analysis would encourage full use of the exergy method. The application of the exergy method to energy-related systems can demonstrate to engineers the benefits that will be gained in systems performance analysis and energy conservation.

Significant in my decision to write this book was the success of the ASME Technical Divisions Conference, "EXERGY – The Efficient Use of Energy", sponsored by the Southern California Advisory Council of the ASME in conjunction with California State University at Los Angeles. This conference was held on May 17, 1975, and attracted a large group of engineers who showed great interest in exergy and in using energy efficiently. This book is based on my presentation "The Exergy Method of Energy System Analysis", that was given at this conference.

I want to thank my colleagues at Aerojet ElectroSystems Company for the interesting discussions about thermodynamics that led me to clarify many aspects of this book. Forrest Cleveland, Dave Curran, Truman Lawson, and Phil Sywulka were especially helpful. I also thank my daughter Maureen, who typed most of the draft manuscript and the translations, and Tim and Jeannine, who performed the calculations for the tables after plugging the exergy equation into the desk programmable calculator.

JOHN E. AHERN

Glendora, California
January 1980

Contents

THE EXERGY METHOD
OF ENERGY SYSTEMS
ANALYSIS

CHAPTER 1
Fundamentals of Energy

The effective use of our energy resources has generally been relegated to secondary importance in favor of initial cost of the installation and the overall operating cost. With increasing cost of our most widely used fuels and the potential decrease in their availability in the not-to-distant future, the importance of the effective use of our available energy resources is now receiving the attention it has always deserved. The location and degree of inefficient use of energy in our energy systems should now be a primary factor in the design and performance analysis of these systems. The *exergy* method of analysis is directed to providing this information by a systematic approach that can be easily added to conventional design and performance calculating procedures.

Higher energy costs shift the economic design criteria from a dominate initial-cost criteria to a greater concern with life-cycle costs. In addition, the possibility of changes in the availability of energy resources as well as their cost dictates that system designs be based on the use of multiple energy sources. According to this concept, switchover to alternate sources can be made readily when the primary energy source is not available or becomes less cost-effective than the alternate.

The use of multiple or combined energy sources can follow several paths. One approach is to design systems to be initially capable of multiple fuel sources similar to that used in the 1950s for power where coal, oil, or gas could be used in a single boiler with minimum changeover. Another approach, which will be more prevalent as time goes on, is to use newly developed low-energy sources such as solar, wind, ocean thermal, and tidal, supplemented by the expendable fuels when the prime energy source is unavailable or inadequate.

In addition to using combined energy sources in our system designs of the future, emphasis is given to the combined use of a single energy source to increase the effective use of the available work in the energy source. This

1

approach is already under way in the designs now being developed to make use of waste heat from gas-turbine plants for steam generation to provide additional power, for space heating, and to provide steam or heat for chemical processing. Topping cycles such as the gas-turbine–steam power plant and bottoming cycles such as the steam–Freon power plant will find application in the effective use of energy in the future. All these system concepts are more complex than our conventional systems and involve direct interaction between discrete cycles that have varying operational characteristics. To assess the effectiveness of energy use in these systems, a realistic measure of energy utilization must be applied. The exergy method of analysis will provide this true measure of effective energy use through its application of principles of both the first and second laws of thermodynamics. The exergy method can also provide a more accurate analysis of energy systems from the overall economic viewpoint. These points are exemplified in subsequent chapters of the book.

The exergy method of analysis is a systematic approach to applying the first and second laws of thermodynamics to components and processes. The building-block approach presented in this book can be applied to any system. It is relatively simple to use, and the benefits gained in more accurate and complete system knowledge far outweigh the additional effort in the analysis.

The fundamentals of the exergy method are embodied in the principles of entropy production and the second law of thermodynamics that are well described in thermodynamic textbooks, many of which are listed in the bibliography at the end of this book. However, the application of the second law of thermo-dynamics has been limited. The reason for the neglect of the second law in engineering systems analysis and design is apparently threefold: (1) the analysis of engineering systems in books and reports generally are based only on the first law heat balance; (2) examples of the second law are generally limited to simple processes or simple cycles where the benefits of using second-law analysis are not apparent; and (3) the design and operating conditions for an energy-related plant have usually been based on initial plant costs with energy costs and energy conservation relegated to secondary consideration.

With the increasing interest in energy conservation and the need to reduce energy costs, the design of energy-related industrial systems becomes increasingly critical and complex. To maximize the use of high-quality energy resources, systems will become more integrated, tending toward combined systems where practical. Irreversible losses in these systems will be a significant factor in the selection of the design and operating criteria.

1-1 PRINCIPLE OF WORK

Our energy resources are now being expended at a significant rate to provide the needs of our advanced industrial society. The final product from the expenditure of these energy resources is work that we use to perform tasks such as moving an automobile or tractor, pumping water to our cities, and lighting our homes. Work is made available from the energy resources in many forms. For example, the combustion of oil or gas in a power plant provides high-pressure, high-temperature steam that is available to do work through a turbine and generator system. The work available in the steam decreases as its temperature and pressure are reduced by flowing through the piping, valves, and turbine of the system. When the temperature and pressure of the steam are near the conditions of the surrounding environment (condensed liquid near ambient temperature), the work available in the steam has essentially disappeared. The water that is held behind a dam on a river is available to do work by flowing it through a hydraulic turbine and an electric generator. The available work in the water behind the dam reverts to zero when the water falls to sea level. Therefore, the steady-state condition of our surrounding environment is a good base to which work can be referenced for all terrestrial systems. It is this concept of available work referenced to the surrounding environment that is the basis of the exergy method of energy-systems analysis.

Although available work can take many forms (e.g. pressures, temperatures, and velocities) that are different from the ambient surroundings, they can all be directly related to each other through conversion factors and the mechanical equivalent of heat. Therefore, the use of available work or exergy is a realistic method of comparing the efficient use of our energy resources. It should be noted that a fluid or gas that is not at equilibrium with the ambient surroundings has the potential to perform work as its condition reverts to the ambient surrounding conditions, as everything will do naturally. This means that a fluid that is colder than the ambient surroundings will be available to perform work as it warms up to the ambient surroundings just as a warm fluid is available to perform work in its passage to the ambient surrounding conditions.

The way we use the available work in our energy resources is a very important factor in our energy-conservation efforts. If we do not make effective use of the work that is available, we will more rapidly deplete our energy resources while accomplishing the same tasks. As the available work in a system working fluid decreases through energy-related processes, there are losses in the available work since no transfer of heat or conversion between mechanical work and heat can be performed without some irreversibility in the process. In a system in which many processes are involved, the loss of work in the system will be distributed throughout the individual processes. It is important to establish the relative losses in each process if we are to effectively improve the system efficiency. Only with this information will we be able to know where to concentrate our energy-conservation efforts.

It should be noted that the conventional heat-balance method of evaluating system losses and system efficiency is misleading and not a true representation of system effectiveness. Only through an evaluation of the available work throughout the system can we have a true measure of the losses in the system processes, which is necessary for effective energy conservation in system design and operation.

1-2   RELATION OF THERMODYNAMICS TO ENERGY

The conversion between work and heat falls in the science of thermodynamics. The transfer of heat to work, and vice versa, must be done within the limitations imposed by the two laws of thermodynamics. Since most of our energy systems involve heat and either mechanical or electrical work, thermodynamics plays an important role in the design and operation of industrial plants that either generate or use energy.

In systems involving the use of both heat and electrical or mechanical work, the conversion between heat and work results in significant losses in available work because of the irreversibilities involved in the processes. Even the conversion from chemical energy to heat is a relatively inefficient process when viewed from the standpoint of loss in available work that occurs in the process.

There are a few energy systems that do not involve thermodynamic principles. However, in many mechanical or electrical systems it is possible to neglect the thermodynamic effects since they are of secondary importance in the overall system design and operation. Examples of this latter type include piping systems where electrical power is used for the pump drive. The performance and efficiency of this system can be completely evaluated by mechanical and electrical work principles if the pumped fluid does not require heating, as would be the case with heavy oil. Another example is the electrical transmission system where Joule heating in transformers and lines is so small as to be negligible in the overall energy balance of the system.

The primary role that thermodynamics plays in energy systems is through the first and second laws of thermodynamics. The first law of thermodynamics, discussed in a later section, is widely used in engineering practice and is the basis of the heat-balance method of analysis that is commonly used in energy-systems performance analysis. The second law of thermodynamics involves the reversibility and irreversibility of processes and is a very important aspect in the exergy method of energy-systems analysis. The second law of thermodynamics is covered well in most thermodynamics textbooks, but its application in engineering practice is limited. The exergy method of analysis as described in this book provides a practical and effective method of applying the second law of thermodynamics throughout a complete system.

1-3 THERMODYNAMIC PROCESSES AND CYCLES

Energy systems are made up of a series of individual processes that form closed or open cycles. Each process in a system or cycle can be analyzed separately from the system by performing a first-law energy balance around the component involved in the process. Figure 1-1 shows an energy balance around a gas compressor. The performance of the process is described best by showing it on pressure—volume and temperature—entropy diagrams. The common thermodynamic processes for gases are shown in Figure 1-2, and the equations associated with these processes are given in Table 1-1. These equations are used in engineering practice to evaluate the change in conditions of a working fluid that are induced by the process. These equations, along with the energy balance (described later), are used to calculate the state of the working gas in any cycle or system. Where the working fluid is a liquid that undergoes phase changes in the cycle (boiling or condensing), the state of the fluid can be developed from properties, tables, and the energy balance.

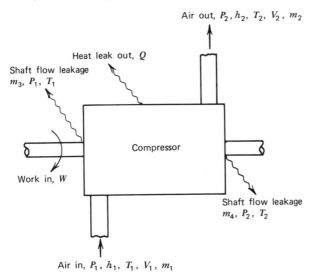

Figure 1-1. Energy balance of a system component.

A cycle is a series of processes and may be open, closed, or interconnected (Figure 1-3). All cycles have an interface with the ambient surrounding or with another cycle, and each type of cycle treats this interface differently. In a closed cycle the working fluid remains within the closed system, and the interface with the surrounding is at boundaries through which heat or work is transferred. The conventional household refrigerator and the steam system of a con-

Gas process

Constant—volume process

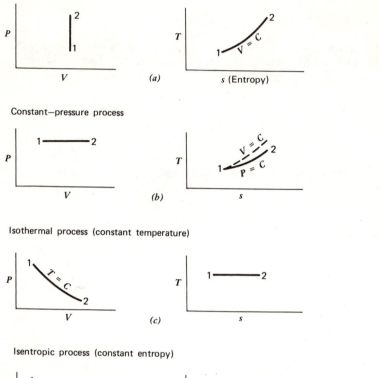

(a)

Constant—pressure process

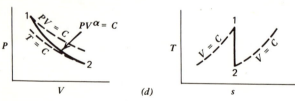

(b)

Isothermal process (constant temperature)

(c)

Isentropic process (constant entropy)

(d)

Polytropic process

The slope can be any value, depending on values of n. In practice n does not vary much from γ or 1.40.

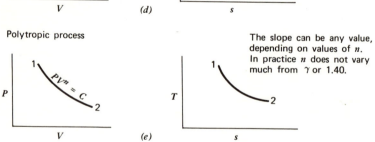

(e)

Figure 1-2. Common thermodynamic processes: (*a*) constant-volume process; (*b*) constant-pressure process; (*c*) isothermal; (*d*) isentropic; (*e*) polytropic.

6

Table 1-1   Perfect Gas Equations

Symbols	Perfect Gas Relations	Work in Nonflow Expansion or Compression	Change in Internal Energy	Heat Transfer to and from System	Change in Enthalpy	Change in Entropy
Symbols	p, v, T	$\int_1^2 p\,dv$	$u_2 - u_1$	Q	$h_2 - h_1$	$s_2 - s_1$
Constant volume $(v = c)$	$\dfrac{T_2}{T_1} = \dfrac{p_2}{p_1}$	0	$wc_v(T_2 - T_1)$	$wc_v(T_2 - T_1)$	$wc_p(T_2 - T_1)$	$wc_v \exp \dfrac{T_2}{T_1}$
Constant pressure $(p = c)$	$\dfrac{T_2}{T_1} = \dfrac{v_2}{v_1}$	$p(v_2 - v_1)$	$wc_v(T_2 - T_1)$	$wc_p(T_2 - T_1)$	$wc_p(T_2 - T_1)$	$wc_p \exp \dfrac{T_2}{T_1}$
Constant temperature $(T = c)$	$p_1 v_1 = p_2 v_2$	$p_1 v_1 \exp \dfrac{v_2}{v_1}$	0	$\dfrac{p_1 v_1}{J} \exp \dfrac{v_2}{v_1}$	0	$\dfrac{wR}{J} \exp \dfrac{v_2}{v_1}$
Constant entropy $(s = c)$	$p_1 v_1^k = p_2 v_2^k$ $\dfrac{T_2}{T_1} = \left(\dfrac{p_2}{p_1}\right)^{(k-1)/k}$ $\dfrac{T_2}{T_1} = \left(\dfrac{v_1}{v_2}\right)^{k-1}$	$\dfrac{p_2 v_2 - p_1 v_1}{1-k}$	$wc_v(T_2 - T_1)$	0	$wc_p(T_2 - T_1)$	0
Polytropic $(pv^n = c)$	$p_1 v_1^n = p_2 v_2^n$ $\dfrac{T_2}{T_1} = \left(\dfrac{p_2}{p_1}\right)^{(n-1)/k}$ $\dfrac{T_2}{T_1} = \left(\dfrac{v_1}{v_2}\right)^{n-1}$	$\dfrac{p_2 v_2 - p_1 v_1}{1-n}$	$wc_v(T_2 - T_1)$	$Wc_n(T_2 - T_1)$	$wc_p(T_2 - T_1)$	$wc_n \exp \dfrac{T_2}{T_1}$

ventional power plant are examples of the closed cycle in which the working fluid remains within the system.

The open cycle makes use of surrounding atmosphere to close the cycle or as the working fluid. The aircraft jet engine is an example of the open cycle where the air is drawn into the engine and processed by entering into combustion to provide work before being rejected back to the atmosphere. A system that liquefies air to produce oxygen for steel making can also be considered an open cycle since air is introduced into the cycle and enters into the series of processes required to produce the liquid oxygen.

There are other systems that combine open and closed cycles in different ways to perform tasks more economically. One system now under serious consideration is the combined cycle shown in Figure 1-3, where the exhaust from a gas-turbine cycle is used in a steam or Freon closed cycle to provide additional work. Another practical combined cycle system is the cogeneration of power and heat for domestic heating or processing in the same system. This latter system is very common in the Soviet Union.

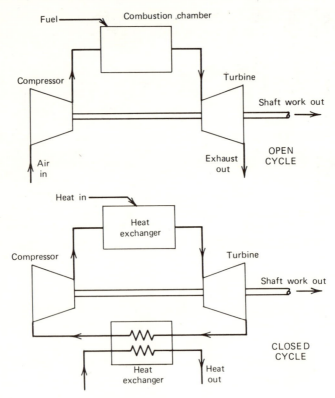

Figure 1-3.   Types of energy cycle.

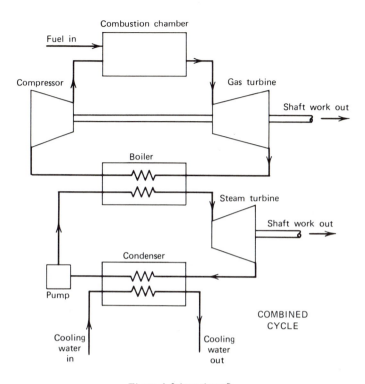

Figure 1-3 (*continued*)

There are several types of cycle used for power and refrigeration; selection depends on many factors, including initial cost, efficiency, operating costs, and compatability with the requirements and the environment. The cycles as described here are ideal, and under actual operating conditions the ideal processes are modified by inefficiencies resulting from friction, heat loss, and the inability of machine devices to achieve the ideal process.

Heat and Power Cycles

Carnot Cycle. The conventional Carnot cycle shown in Figure 1-4 is the most efficient cycle between any two temperatures. It consists of the following processes:

1—2   Isothermal process — heat flows from heat source to compressor at constant temperature.

2—3   Isentropic process — entropy constant and temperature drops since work is being done at the expense of internal energy.

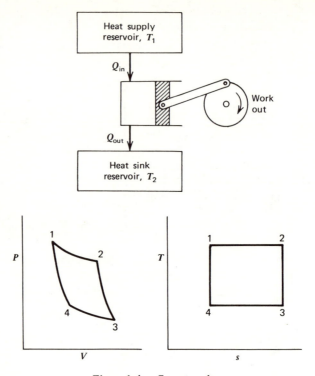

Figure 1-4. Carnot cycle.

3–4   Isothermal process – heat flows from compressor to heat sink at constant temperature.

4–1   Isentropic process – entropy constant and temperature increase since compression increases the internal energy.

The efficiency of the Carnot cycle is

$$\frac{Q_A - Q_R}{Q_A} = \frac{(T_1 - T_2)(s_2 - s_1)}{T_1(s_2 - s_1)} = \frac{T_1 - T_2}{T_1}$$

where s_1 = entropy at 1
   s_2 = entropy at 2
   Q_A = heat added at T_1
   Q_R = heat rejected at T_2

The Carnot cycle is not directly used in actual energy systems, but it does approximate some refrigeration systems. It provides the efficiency reference for other cycles and systems since it is the ideal energy cycle between any two temperatures.

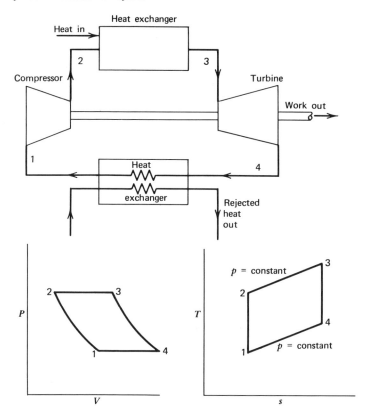

Figure 1-5. Brayton cycles; (*a*) closed cycle (*above*); (*b*) basic open cycle (*page 12*); (*c*) open cycle with recuperative heat exchanger (*page 13*).

Brayton Cycle.   The Brayton cycle (Figure 1-5) uses a gas as the working fluid. This cycle consists of the following processes:

1–2   Isentropic compression of the gas in the compressor.

2–3   Constant-pressure heat addition in the combustion chamber.

3–4   Isentropic expansion in the turbine where work is done.

4–1   In the closed cycle the gas is returned to the compressor inlet usually after being cooled by the heat exchangers. The surrounding air closes the loop in the open cycle.

The closed Brayton Cycle (Figure 1-5a) is used with helium or other gases in some nuclear power-plant designs. The basic open Brayton cycle (Figure 1-5b) is the cycle used in the turboprop engine, where some of the output is also available in the high-velocity exhaust flow from the turbine outlet. The open Brayton cycle is also used for stationary power plants, but a regenerative heat

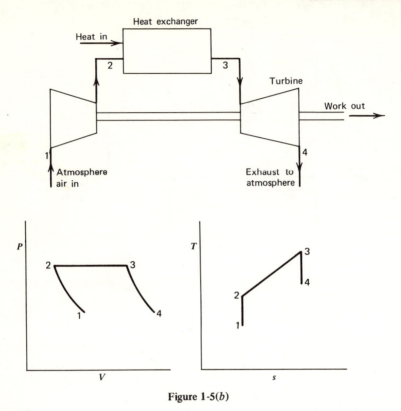

Figure 1-5(b)

exchanger is located between the turbine exhaust and the compressor inlet (Figure 1-5c) to reduce heat-rejection losses and to improve system efficiency.

Rankine Cycle.   The Rankine cycle (Figure 1-6) consists of the following processes:

1–2 Constant-volume pressure increase during pumping of the liquid plus a variable-temperature heat addition to the boiling temperature.

2–3 A constant-pressure heat addition that boils the working fluid at constant temperature from 2 to 3. During additional heating beyond the saturated vapor curve on the $T–s$ diagram the fluid is superheated in a constant-pressure process with the temperature increasing to $3'$.

3–4 This is an isentropic expansion in a turbine or engine to point 4, which is a liquid–vapor mixture at the pressure of point 1. When the vapor is superheated to point $3'$, the expansion in the turbine may terminate in the superheated region at the pressure of point 1 or in the liquid–vapor mixture region.

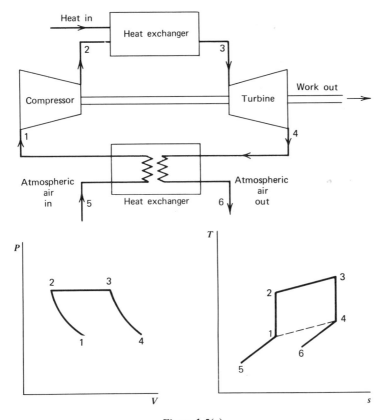

Figure 1-5(c)

4–1   This is a constant pressure condensing process that is at constant pressure in the region where liquid–vapor conditions exist. When the fluid at 4 is in the superheated region, precooling of the vapor is required prior to start of condensation.

The Rankine cycle is used in the common steam power plant and is also being considered with steam and other fluids for solar-powered electric power plants.

Otto Cycle.   This is the open-air cycle of the conventional internal combustion engine and is comprised of the following processes (Figure 1-7):

1–2   Isentropic compression of air from standard atmospheric conditions at point 1.

2–3   Constant-volume heat addition (this is the combustion process, which raises the pressure from point 2 to point 3).

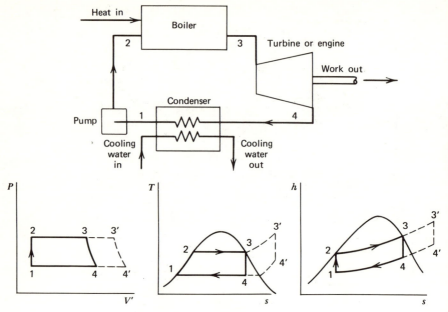

Figure 1-6. Rankine cycle.

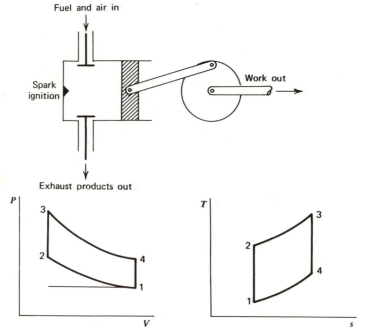

Figure 1-7. Otto cycle.

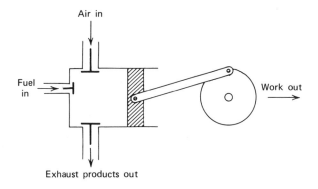

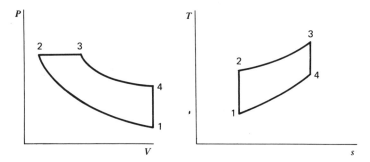

Figure 1-8. Diesel cycle.

3–4   Isentropic expansion in the engine, which is the power stroke.

4–1   Constant-volume exhaust of the combustion products with pressure reduction to atmospheric pressure.

Diesel Cycle. The Diesel (Figure 1-8) is also used in internal combustion engines, but the heat addition is performed in a constant-pressure process instead of the constant-volume process of the Otto cycle. The Diesel cycle is comprised of the following processes:

1–2   Isentropic compression of air from standard atmospheric conditions at point 1.

2–3   Constant-pressure heat addition corresponding to the combustion process while the piston is withdrawing from the cylinder.

3–4   Isentropic expansion where work is performed by the piston.

4–1   Constant-volume exhaust of the combustion products with pressure reduction to atmosphere pressure.

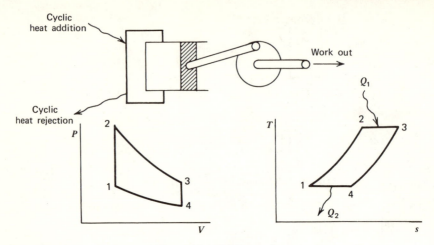

Figure 1-9.   Stirling cycle.

Stirling Cycle.   The Stirling cycle (Figure 1-9) and the Brayton cycle are potential competitors of the Otto and Diesel cycles for automobile engines of the future. The Stirling cycle is the basis of an external combustion closed-cycle engine that can achieve high efficiencies if used with a regenerative heat exchanger within the cycle. The basic Stirling cycle is made up of the following processes:

1–2   Constant-volume heat addition.

2–3   A constant-temperature expansion where heat is added during the expansion stroke of the piston.

3–4   Constant-volume heat rejection.

4–1   Constant-temperature compression where heat is rejected during the compression stroke of the piston.

Before closing this section on heat and power cycles, it should be noted that there are many common variations of these basic cycles. Some of these variations, such as reheat in steam cycles and intercooling in gas-turbine Brayton cycles, are important features in achieving greater overall system efficiency.

Refrigeration Cycles.   Many of the heat and power cycles described previously are used in the reverse direction to provide refrigeration from the power supplied to the cycle. The more common refrigeration cycles are described here.

Reverse Carnot Cycle.   The reverse Carnot cycle (Figure 1-10) is the ideal refrigeration cycle and has the greatest efficiency for refrigeration between any two temperatures. The efficiency of a refrigeration system is generally given in terms of its coefficient of performance (COP), which for the Carnot cycle is

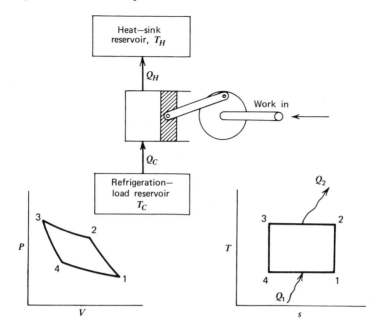

Figure 1-10. Reverse Carnot cycle.

$$\text{COP} = \frac{\text{Refrigeration load}}{\text{Work in}} = \frac{Q_1}{W} = \frac{Q_1}{Q_2 - Q_1} = \frac{T_c}{T_h - T_c}$$

The reverse Carnot cycle is often used as the reference for comparing the relative performance of actual refrigeration systems. Figure 1-11 shows the maximum value of the refrigeration work-input ratio as a function of temperature. Efficiencies higher than this curve cannot be achieved for any refrigeration system at the specific refrigeration temperature.

The Carnot refrigeration cycle is approximated by refrigeration systems in which the refrigerant is vaporized and condensed within the cycle since part of these processes involve constant-temperature heat exchanges. However, because of irreversible losses and the inability to achieve isentropic expansion and compression with the liquid–vapor mixture, actual refrigeration cycles using evaporation and condensation of the refrigerant are closer to the reverse Rankine cycle.

Reverse Brayton Cycle.   The reverse Brayton cycle (Figure 1-12) is used as an open cycle with atmospheric air to provide air cooling. In the closed cycle using helium as the working gas, the Brayton cycle is the basic element in many cryogenic liquefying systems when it is modified with regenerative heat exchangers and combined with other systems such as the Joule–Thomson

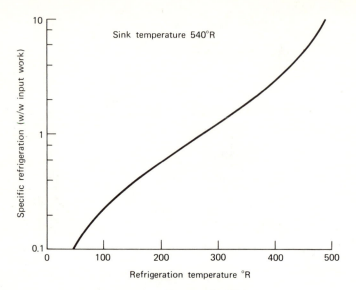

Figure 1-11.   Carnot cycle efficiency.

(J–T) expansion system. As shown in Figure 1-12, the reverse Brayton cycle consists of an isentropic compression, a constant-pressure heat rejection, and an isentropic expansion. In the closed cycle there is an additional constant-pressure heat-addition process performing the refrigeration.

Reverse Rankine Cycle.   The reverse Rankine refrigeration cycle (Figure 1-13) is the basis for the common household refrigerator system and most home air-conditioning systems. The refrigerant is isentropically compressed to a high pressure, where it is cooled and condensed in a heat exchanger at constant pressure. The liquid is then isenthalpically expanded in a throttle valve to a low temperature and pressure. The refrigerant is then evaporated in a heat exchanger by the heat absorbed in performing the refrigeration. Common refrigerants used in systems with the Rankine cycle usually are the Freons, ammonia, methyl chloride, sulfur dioxide, and similar compounds.

Reverse Stirling Cycle.   The reverse Stirling cycle (Figure 1-14) with a regenerator has a constant-volume heat-addition process followed by a constant-temperature compression with heat rejection. There is then a constant-volume heat-removal process followed by a constant-temperature expansion in which heat is added. Because of the inability of machines to approach the constant-temperature processes, the basic Stirling cycle is not practical. However, with the addition of regenerative heat exchangers in the system, as shown in Figure 1-14,

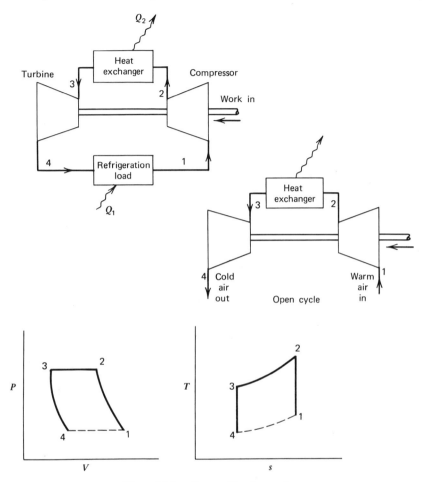

Figure 1-12. Reverse Brayton cycle.

the reverse Stirling cycle is a practical and efficient device for achieving low cryogenic temperatures.

Joule–Thomson Cycle. The J–T refrigeration system (Figure 1-15) makes use of the throttling process to provide a refrigeration effect. The throttling process involves flowing the gas through a restriction with results in an adiabatic expansion of the gas. The expansion results in a change of temperature in an isenthalpic process. The change in temperature with pressure is defined as the J–T coefficient

$$u = \left| \frac{dT}{dP} \right|_h$$

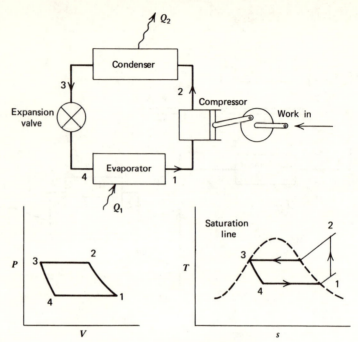

Figure 1-13.   Reverse Rankine cycle.

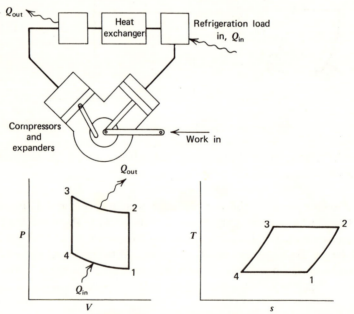

Figure 1-14.   Reverse Stirling cycle.

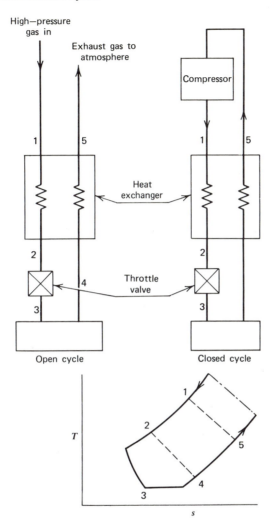

Figure 1-15. Joule–Thomson refrigeration cycle.

which is zero for an ideal gas. For a real gas, the J–T coefficient is zero along a given temperature–pressure line called the *inversion curve*. During adiabatic expansion (throttling) the gas temperature will decrease if the initial gas temperature is below the inversion curve. When the initial gas temperature is above the inversion curve, throttling of the gas will result in a temperature increase. The inversion curves for several fluids used in J–T refrigerators are given in Figure 1-16.

The J–T refrigeration cycle is relatively inefficient because of the complete

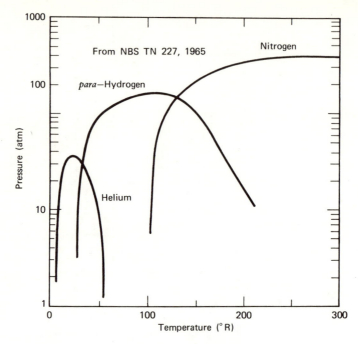

Figure 1-16. Joule–Thomson inversion curves.

loss of available work in the gas during the isenthalpic process. It is replaced by the turbine or piston expander, where some work is recovered from the expansion process. The J–T refrigerator is used in many special applications and in combination with other refrigeration cycles. For example, the J–T process is used as the final step to liquefy helium and other refrigerants.

Vuilleumier Cycle.   The Vuilleumier cycle uses heat to provide refrigeration. This cycle (Figure 1-17) is similar to the Stirling cycle but uses thermal compression instead of the piston compression of the Stirling cycle. Thermal compression is accomplished in the VM refrigerator by the reciprocating power displacer inside the cylinder. The heated gas in the end of the cylinder is alternately increased and decreased in volume by moving it through the regenerator to the crankcase. Since the whole refrigerator is filled with a constant volume of gas, the average gas temperature throughout the refrigerator will be highest when the hot gas volume in the power cylinder is greatest. Consequently, the pressure of the gas throughout the refrigerator will be highest under these conditions. The reverse is true for a low volume in the power cylinder. The refrigeration cylinder of the VM refrigerator is similar to that of the Stirling cycle refrigerator.

The cycle process in the ideal VM refrigerator is presented in Figure 1-17 for

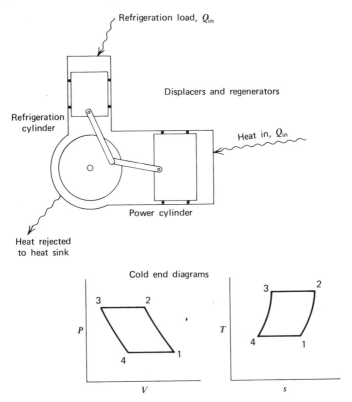

Figure 1-17. Vuilleumier cryogenic refrigerator.

the cold-end volume. The actual VM refrigerator operates more on a polytropic than an isenthalpic process.

Since the VM is a refrigerator driven by a heat engine, its overal theoretical efficiency is the product of the two efficiencies

$$\eta = \frac{Q_c}{Q_h} = \underbrace{\left(\frac{T_h - T_a}{T_h}\right)}_{\substack{\text{Heat} \\ \text{engine}}} \underbrace{\left(\frac{T_c}{T_a - T_c}\right)}_{\substack{\text{Refrig-} \\ \text{erator}}}$$

where T_a is the temperature of the crankcase from which heat is rejected outside the cycle.

Absorption Refrigeration Cycle. The absorption refrigerator cycle (Figure 1-18) operates in a fashion similar to those of other refrigeration cycles, but the compression is accomplished by liquid pumping and the relative concentration

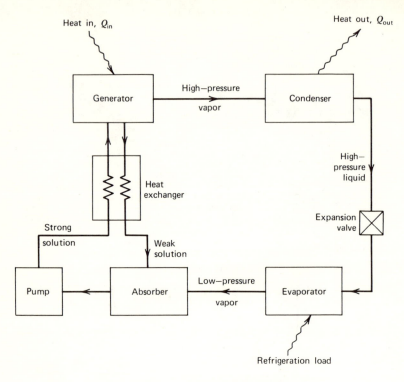

Figure 1-18. Absorption refrigeration cycle.

of the working fluid rather than by mechanical or thermal compression. Since only a small amount of power is required for the liquid pump and low-quality heat (low temperature) is used for the compression, this cycle is well suited for a solar powered refrigerator. The common working fluids for this cycle are ammonia and lithium bromide. Baum et al. (1972) report on the performance of a solar absorption cooling plant using lithium chloride as the working fluid.

1-4   FIRST LAW OF THERMODYNAMICS AND ENERGY BALANCE

In the design and performance analysis of systems involving energy generation or use, the first law of thermodynamics is used to determine the condition of the working fluid throughout the system. This law states that energy is conserved throughout a process or system; therefore, an energy balance around a process or system can be made to account for all the energy. The principle of conservation of mass in a flow process or system is also applied during an energy

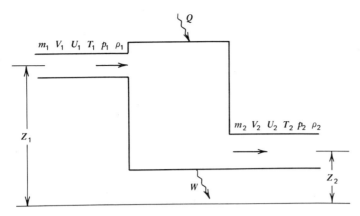

Figure 1-19.   General energy-balance diagram.

balance to show accountability for mass flow leakage from the process or device
that can account for energy loss.

The general-energy diagram (Figure 1-19) indicates the mass flows and energy
quantities going in or out of a device. In the design of power-plant equipment,
engines, and chemical processing equipment, such energy balances are made on
each piece of equipment or process to ascertain conditions of the materials in
the process and other data.

Equating the quantity of energy entering and leaving in Figure 1-19 will
give the general energy equation for m pounds of fluid.

$$\frac{gz_1}{J} + \frac{V_1^2}{2gJ} + u_1 + \frac{p_1 v_1}{J} + Q$$

$$\begin{array}{ccccc}
\text{Potential} & \text{Kinetic} & \text{Internal} & \text{Flow} & \text{Work} \\
\text{energy} & \text{energy} & \text{energy} & \text{work} & \text{in} \\
\text{in} & \text{in} & \text{in} & \text{in} &
\end{array}$$

$$= \frac{gz_2}{J} + \frac{V_2^2}{2gJ} + u_2 + \frac{p_2 v_2}{J} + W$$

$$\begin{array}{ccccc}
\text{Potential} & \text{Kinetic} & \text{Internal} & \text{Flow} & \text{Work} \\
\text{energy} & \text{energy} & \text{energy} & \text{work} & \text{out} \\
\text{out} & \text{out} & \text{out} & \text{out} &
\end{array}$$

$$(1\text{-}1)$$

The unit for all the preceding terms is the British thermal unit (Btu), and J
is the Joule constant and is the equivalent number of foot-pounds of mechanical
energy for one Btu of heat energy. Its value is 778. In any specific application,
one or more of the terms may be zero or negligible. The internal energy term, u,
and the flow-work term, pv, can be combined into an enthalpy, h, term since
$h = (u + pv)$.

The following examples illustrate the use of the first law of thermodynamics on a process.

Example 1-1   An air compressor is used to supply 1000 cfm (cubic feet per minute) of ambient air to an ejector nozzle at 200 psia. Leakage from the high-pressure side of the compressor to the low-pressure side has been calculated as 50 cfm. The inlet air velocity is essentially negligible, and the outlet velocity is 50 ft/sec. Determine the power required by this ideal compressor from an energy balance around the compressor unit. Figure 1-20 illustrates the mass and

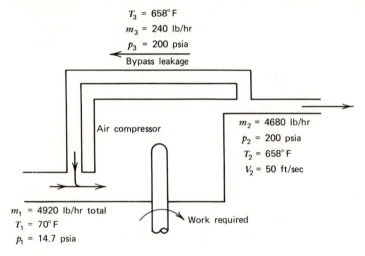

$T_3 = 658°$ F
$m_3 = 240$ lb/hr
$p_3 = 200$ psia
Bypass leakage

Air compressor

$m_2 = 4680$ lb/hr
$p_2 = 200$ psia
$T_2 = 658°$ F
$V_2 = 50$ ft/sec

$m_1 = 4920$ lb/hr total
$T_1 = 70°$ F
$p_1 = 14.7$ psia

Work required

Figure 1-20.   Mass and energy balance, example 1.1.

energy-balance terms for this process of compression. The mass balance around the compressor is

$$M_1 \quad = \quad M_2 \quad + \quad M_3$$
$$4680 \text{ lb/hr} = 4440 \text{ lb/hr} + 240 \text{ lb/hr}$$

and the energy balance is

$$w + M_1 h_1 = M_2 h_2 + M_3 h_3$$

From air tables, $h_1 = 126.7$ Btu/lb. The data lacking for completing the last balance are the enthalpies h_2 and h_3 and the work input W. Only two of these three values are required to complete the heat balance. From Table 1-1, assuming that the compressor has a 100% isentropic efficiency,

$$\frac{T_{2,3}}{T_1} = \left(\frac{p_{2,3}}{p_1}\right)^{(k-1)/k}$$

where k for air is 1.4 and $T_1 = 530°$R.

Then

$$T_{2,3} = T_1 \left(\frac{200}{14.7}\right)^{0.286} = 1118\,°R$$

$$T_{2,3} = 658\,°F$$

$$\Delta h = 0.24(1118 - 530) = 141.1\,\text{Btu/lb}$$

$$h_2 = h_3 = h_1 + \Delta h = 126.7 + 141.1 = 267.8\,\text{Btu/lb}$$

The heat balance can now be written

$$W + (4920\,\text{lb/hr} \times 126.7\,\text{Btu/lb})$$

$$= \frac{4680\,\text{lb/hr}\,(50\,\text{ft/sec})^2}{2(32.2\,\text{ft/sec}^2)778.2\,\text{ft/lb Btu}}$$

$$+ (4680\,\text{lb/hr} \times 268\,\text{Btu/lb})$$

$$+ (240\,\text{lb/hr} \times 267.5\,\text{Btu/lb})$$

$$W + 623,364\,\text{Btu/hr} = 233\,\text{Btu/hr} + 1,253,304\,\text{Btu/hr} + 64,272\,\text{Btu/hr}$$

$$\text{Work (required)} = 694,445\,\text{Btu/hr}$$

$$= 203\,\text{kW}$$

$$\text{Required work input} = 272\,\text{hp}$$

Under actual conditions it is difficult to calculate realistically the losses associated with leakage, bearing friction, seal wear, internal heat flow, and other variables in complex devices such as compressors and turbines. If no losses existed, the process would be ideal and performance as given by isentropic relationships. Actual performance characteristics are determined from test-block or field operational tests. The difference between the actual and ideal performance is designated as the machine efficiency

$$\eta_{\text{machine}} = \frac{\eta_{\text{actual}}}{\eta_{\text{isentropic}}}$$

The following example illustrates this procedure.

Example 1-2   Determine the machine efficiency of the compressor in Example 1-1 without the bypass leakage or momentum losses if the input power is measured to be 300 hp during field operation. The heat balance under ideal (isentropic) compression is

$$W = (4680\,\text{lb/hr} \times 126.7\,\text{Btu/lb}) = 592,596\,\text{Btu/hr}$$

and the input power required is

$$W = 1{,}253{,}304 - 592{,}956 = 660.348 \text{ Btu/hr}$$

$$= 262 \text{ hp}$$

The actual efficiency is then

$$\eta_{\text{actual}} = \frac{\text{output}}{\text{input}} = \frac{4680 \text{ lb/hr} \times 141.1 \text{ Btu/lb}}{300 \text{ hp} (2545 \text{ Btu/hp})} = \frac{660{,}348}{763{,}500}$$

$$\eta_{\text{actual}} = 0.86$$

Since the isentropic efficiency was assumed to be 100%, the compressor (machine) efficiency is then

$$\eta_{\text{compressor}} = \frac{\eta_{\text{actual}}}{\eta_{\text{isentropic}}} = \frac{0.86}{1.00} = 0.86$$

The energy balance [(Eq. (1-1)] is used in energy-systems analysis to calculate the state of the working fluid at the stations in the system. Starting at a known or assumed point in the system, the condition of the working fluid is calculated at all other stations in the system by making an energy balance around each process or component.

For consistency in the analysis and evaluation of cycles and systems, steady-state conditions will be assumed in all cases. The flow work required by a process or introduced into the system by a process will be included as part of the specific process. For example, the loss of available work because of flow frictional losses in a heat exchanger will be included as part of the heat-exchanger effectiveness, even though the work to overcome this frictional loss is provided by a compressor or pump in another process in the system. The reason for assuming steady-state conditions will be apparent in the exergy analysis of more complex systems where the accountability of the loss in available work from the energy resource is of primary importance.

Example 1-3 illustrates the use of the first-law heat balance in a complete engine cycle.

Example 1-3   To illustrate the use of the isentropic relationships and to show how basic cycle calculations are made, the following example is presented. Air at 14.7 psia initial pressure enters the compressor of a gas-turbine power plant. The compressor pressure ratio is 6. The specific heats of the air are considered constant $c_p = 0.240$ and $c_v = 0.1715$, and γ is 1.40. The initial air temperature is $40\,^{\circ}\text{F}$ and the temperature entering the turbine, $1200\,^{\circ}\text{F}$ or $1660\,^{\circ}\text{R}$. The Brayton cycle is applicable here.

Given $P_1 = 14.7$ psia, $T_1 = 40\,^{\circ}\text{F}$, or $500\,^{\circ}\text{R}$,

$$\frac{T_2}{T_1} = \left(\frac{P_2}{P_1}\right)^{(\gamma-1)/\gamma} \qquad \therefore \quad \frac{T_2}{T_1} = (6)^{(\gamma-1)/\gamma} \qquad \gamma = 1.40$$

$$\frac{T_2}{T_1} = (6)^{[(1.4)-1]/1.4} = 6^{0.286} = 1.67 \qquad T_2 = 1.67 \times 500 = 836\,^\circ R$$

The specific volume of air at 14.7 psia and 40 °F is 12.50 ft³lb⁻¹.

$$\frac{P_2}{P_1} = \frac{V_1}{V_2} = \left(\frac{P_2}{P_1}\right)^{1/\gamma} \qquad \frac{V_1}{V_2} = (6)^{1/1.4} = 6^{0.715} = 3.6$$

$$V_2 = \frac{12.5}{13.6} = 3.48 \qquad P_2 = 6P_1 = 6(14.7) = 88.3\,\text{psia}$$

The process from 2 to 3 is at constant pressure, so $P_3 = 88.3$ psia and T_3 is given as 1660 °R.

Since the turbine exhausts to atmosphere, $p_4 = 14.7$ psia. Then, using the foregoing procedure with $(p_4/p_3) = 0.167$,

$$\frac{T_4}{T_3} = (0.167)^{0.286} = 0.60$$

$$T_4 = 0.6(1660) = 995\,^\circ R$$

The specific volume at 3

$$V_3 = \frac{T_3}{T_2}$$

since the heat addition from 2 to 3 is at constant pressure.

$$V_3 = 3.48\left(\frac{1660}{836}\right) = 6.28$$

We can now determine the heat transferred in each process by

$$Q = WC_p\,\Delta T$$

To simplify calculation, a unit weight flow is used so that $W = 1.0$. From 1-2,

$$\text{Work of compression} = C_p(T_2 - T_1) = 0.24(836 - 500) = 80.6\,\text{Btu}$$

From 2-3,

$$\text{Heat added in combustion} = 0.24(T_3 - T_2) = 0.24(1660 - 836) = 198\,\text{Btu}$$

From 3-4,

$$\text{Work to turbine} = 0.24(T_3 - T_4) = 0.24(1660 - 995) = 159.6\,\text{Btu}$$

From 4-1,

$$\text{Heat lost in exhaust} = 0.24(T_4 - T_1) = 0.24(995 - 500) = 120\,\text{Btu}$$

Net work or output $= 159.6 - 120 = 79.6$ Btu

Actual efficiency of cycle $= \dfrac{\text{net work}}{\text{heat added}} = \dfrac{79.6}{159.6} = 0.398$

Airflow $= \dfrac{2545}{79} = 32.2$ lb/hp hr

Btu input $= \dfrac{2545}{0.398} = 6400$ Btu/hp hr

The preceding examples illustrate the importance of the first law of thermodynamics and the heat balance in the design and performance analysis of energy processes and systems. However, the first law and the heat-balance analysis provide only the theoretical (ideal) performance or the actual (total) performance of the process or system. To assure that new or modified processes and systems are energy efficient, it is necessary to perform more detailed analyses to establish the magnitude, location, and type of loss involved in the process or system. The second law of thermodynamics is a factor in this latter analysis.

1-5   SECOND LAW OF THERMODYNAMICS AND PRODUCTION OF ENTROPY

Ideal processes that are reversible do not occur in the real world, and the transfer of heat or the conversion of heat and work from one form to another always results in some work loss. The fact that these losses occur in every energy process to some degree means that each time the working fluid in a system goes through a process, some of the initial available work in the working fluid is lost. Because of this loss, the working fluid cannot be returned to its initial state prior to the process without the aid of external work to account for the lost work. Under these conditions the process is irreversible. The irreversibility of a process is accounted for in the second law of thermodynamics and is indicated by an increase in entropy production that is not matched by an equivalent production of work.

The second law of thermodynamics points out that the most energy-efficient closed cycle to perform the conversion of heat to work or work to heat is the Carnot cycle. However, this maximum efficiency is only available under ideal performance conditions shown for the reversible Carnot cycle in Figure 1-21. In this cycle some of the heat added at T_1 must be rejected to the surrounding atmosphere at T_0. Since a temperature gradient is required for heat to flow, the heat from the Carnot cycle must be rejected at some temperature T_2 that is above the surrounding ambient temperature T_0. The input heat Q_1 is represented by the area under the curve 1–2

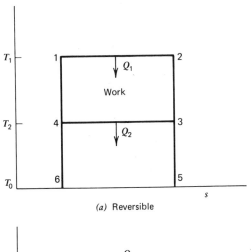

(a) Reversible

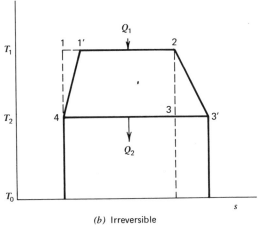

(b) Irreversible

Figure 1-21. $T-s$ Diagrams for Carnot cycle: (a) reversible cycle; (b) irreversible cycle.

$$Q_1 = T_1(s_2 - s_1)$$

The rejected heat Q_2 is represented by the area under the curve 3–4

$$Q_2 = T_2(s_4 - s_3)$$

The net work is

$$Q_1 - Q_2 = T_1(s_2 - s_1) - T_2(s_4 - s_3)$$

However, for the reversible Carnot cycle with isentropic expansion and compression processes, $s_4 - s_3$ is equal to $s_2 - s_1$. Therefore, the net work is

$$Q_1 - Q_2 = (T_1 - T_2)(s_2 - s_1)$$

The efficiency is then

$$\eta = \frac{\text{Net work}}{\text{Heat in}} = \frac{Q_1 - Q_2}{Q_1} = \frac{(T_1 - T_2)(s_2 - s_1)}{T_1(s_2 - s_1)} = \frac{T_1 - T_2}{T_1} = 1 - \frac{T_2}{T_1}$$

Example 1-4   If a closed-cycle engine receives heat from an infinite reservoir at $1500\,°R$ and rejects heat at $600\,°R$, what is the maximum efficiency, at which this cycle can operate?

$$\eta_{\text{ideal, Carnot}} = 1 - \frac{T_2}{T_1} = 1 - \frac{600}{1500} = 0.60$$

This shows that even under ideal conditions where the system operates with no losses, the engine cannot convert more than 60% of the available energy in the input heat into useful work.

The fact that even under ideal conditions a closed-cycle engine cannot achieve 100% efficiency has resulted in two basic types of efficiency. One efficiency is an overall engine or system efficiency that is the actual net work divided by the total energy input to the system. This efficiency includes the Carnot efficiency. The other type of efficiency is called the *engine efficiency* and is

$$\eta_{\text{engine}} = \frac{\text{Work out/energy in}}{\eta_{\text{ideal, Carnot}}}$$

In the ideal (reversible) Carnot cycle (Figure 1-21a) the expansion and compression processes 2–3 and 4–1 are isentropic, and the entropy increase during heat addition $s_2 - s_1$ is equal to the entropy decrease $s_4 - s_3$. Therefore, this ideal cycle has a net entropy change of zero. In an actual (irreversible) Carnot cycle (Figure 1-21b) the compression process is nonisentropic, as shown by line 4–1' because of mechanical, fluid-flow, and heat loss during the actual compression process. Note that this nonisentropic process results in an increase in the entropy s of $s_1' - s_4$ or $s_1' - s_1$. The nonisentropic expansion process from 2–3' also results in an increase in entropy of $s_3' - s_2$ or $s_3' - s_3$.

The consequence of nonisentropic performance for the irreversible cycle is that there is a net increase in the production of entropy for the cycle. This production of entropy is the measure of the irreversibility in a process or cycle. A realistic design approach for systems is to base the design on minimum entropy production.

Another source of entropy production in energy systems is shown in Figure 1-22, where the irreversibility resulting from heat transfer across a finite-temperature difference is considered. The heat transfer from the heat source at T_1' to the cycle at T_1 is accompanied by an increase in entropy as a result of the irreversibility of the finite-temperature difference. The amount of entropy production induced by this heat-transfer process is

$$\Delta s = (s_4 - s_3) - (s_2 - s_1) > 0$$

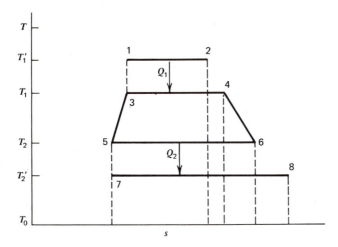

Figure 1-22. Irreversible heat-transfer effect.

The same situation occurs when heat is transferred from the cycle at T_2 to the heat sink at T_2'. In this case the entropy produced in the process as a result of irreversibility is

$$\Delta s = (s_8 - s_7) - (s_6 - s_5) > 0$$

These irreversible losses caused by heat transfer across finite-temperature differences have generally been neglected in energy-systems performance analysis and design. However, with the increased emphasis on energy conservation, these losses become important because of the trend to combined cycles and integrated energy-systems operation where heat transfer between cycles and systems is more prevalent.

The primary message we get from the second law of thermodynamics is that entropy production must be minimized for efficient energy conservation. The work lost through entropy production is accounted for in the exergy method described in Section 1-6.

1-6 DEFINITION OF EXERGY

Exergy is defined as the work that is available in a gas, fluid, or mass as a result of its nonequilibrium condition relative to some reference condition. The sea-level, atmospheric condition is the ultimate sink for all terrestrial energy systems. It is taken generally as the reference condition. Thus steam at $1000\,°F$ and 200 psia pressure has a specific quantity of exergy that is dissipated as the temperature and pressure approach atmospheric conditions. Useful work can be recovered during the cooling and expansion processes of the steam through a steam engine or turbine and heat exchangers. The exergy that is not recovered as useful work is lost.

Exergy is an explicit property at steady-state conditions. Its value can be calculated at any point in an energy system from the other properties that are determined from an energy balance on each process in the system. Exergy is calculated at a point in the system relative to the reference condition by the following general equation:

$$\text{Exergy} = (u - u_0) - T_0(s - s_0) + \frac{P_0}{J}(v - v_0) + \frac{V^2}{2gJ} + (z - z_0)\frac{g}{g_c J}$$

          Internal   Entropy   Work Momentum Gravity
          energy

$$+ \sum_c (\mu_c - \mu_0)N_c + E_i A_i F_i (3T^4 - T_0^4 - 4T_0 T_3) + \ldots$$

     Chemical   radiation emission   (1-2)

where the subscript 0 denotes the reference condition. There are variations of this general exergy equation, and in most systems analysis some, but not all, of the terms shown in Eq. (1-2) would be used. Since exergy is work available from any source, terms can be developed using electrical current flow, magnetic fields, and diffusional flow of materials. One common simplification is to substitute enthalpy for the internal energy and Pv work terms that is applicable for steady flow systems.

The general exergy equation is often used under conditions where the gravitational and momentum terms can be neglected. In this case the equation reduces to

$$Ex_m = (h_m - h_0) - T_0(s_m - s_0) \qquad\qquad (1\text{-}3)$$

This equation can be used to develop the value of exergy diagrammatically. To do this, a reference line must be developed over the enthalpy and entropy range that has a value of exergy equal to zero. This line is represented by Eq. (1-3) for all conditions when $Ex_m = 0$. In this case

$$(h_m - h_0) - T_0(s_m - s_0) = 0$$

Letting dh be the change in enthalpy and ds the change in entropy, we get

$$dh - T_0\, ds = 0$$

Dividing through by T_0, ds gives

$$\frac{dh}{T_0\, ds} = 1$$

or

$$T_0 = \frac{dh}{ds}$$

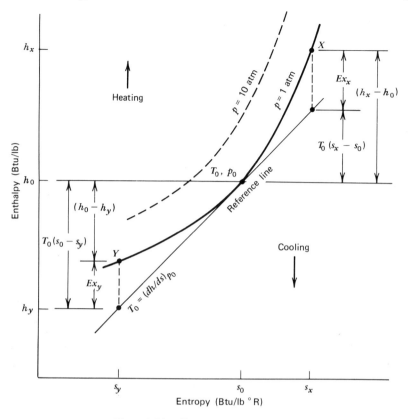

Figure 1-23. Exergy diagram for gases.

which also has the condition of a constant pressure, so the final equation for this reference line is

$$T_0 = \left(\frac{dh}{ds}\right)_{p_0} \tag{1-4}$$

Figure 1-23 shows this reference line for illustrating exergy values of gases in heating and cooling systems. This line is drawn tangential to the 1-atm line at T_0 and is represented by Eq. (1-4). The points x and y are shown located on the 1-atm line, but they could have been selected anywhere in the chart at other pressures. Equation (1-3) can be used directly for the heating part of the curves where the enthalpy and entropy values of the system point are greater than the reference values. For the cooling part of the curves, the system values are smaller than the reference values. One way to handle this is to reverse the values of the terms in Eq. (1-3) as done by Trepp (1961). However, the problem is further

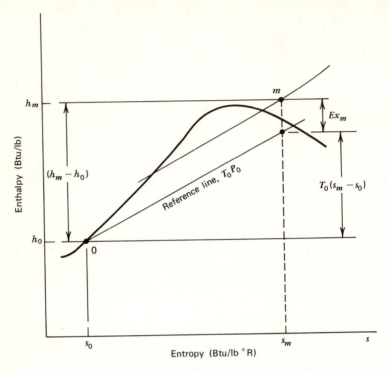

Figure 1-24.   Exergy diagram for liquid–vapor region.

complicated by the fact that the entropy and enthalpy values available in the literature are given for different zero states. Therefore, it is best to consider the absolute values of the terms in Eq. (1-3).

Although exergy values are always positive relative to the reference surrounding environment, the zero reference for enthalpy and entropy values in the literature often results in positive and negative values of exergy being calculated in systems, especially refrigeration systems. This is no problem in a closed-loop system since only the irreversible production of entropy is of interest to develop the internal system exergy losses.

Figure 1-24 shows a special case of the reference line for the liquid–vapor system as encountered in steam power cycles. The constant-pressure line through the liquid–vapor region intersecting the reference point will represent Eq. (1-4).

1-7   RELATION OF EXERGY TO ENERGY

Since exergy is defined as the available work in a mass, it is a special case of defining available energy. The conventional definition of energy as the number

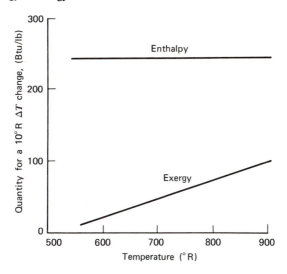

Figure 1-25. Comparison of enthalpy and exergy values as a function of temperature (air at 1 atm).

of British thermal units (Btus) does not consider the quality of the Btus involved; therefore, it is not a true measure of the usefulness of the available energy. One British thermal unit of energy at 3000 °F can perform much more work than 1 Btu at 300 °F. Therefore, consideration of the number of Btus of energy alone is not adequate to determine to effectiveness of our energy-conservation measures. Figure 1-25 illustrate the difference in the values of exergy and enthalpy as a function of temperature.

Examples given later in the text on different energy systems illustrate that the use of exergy is required to show the true quantities and locations of the losses in our available energy resources. This knowledge is important if we are to reduce work losses and raise the effectiveness of using our available energy resources.

CHAPTER 2
Background

The exergy method of analysis is based on the second law of thermodynamics and the concept of irreversible production of entropy. The fundamentals of the exergy method were laid down by Carnot in 1824 and Clausius in 1865. Although all thermodynamic textbooks have chapters on the second law and entropy, the use of the second law has been very limited. Energy-related engineering systems are designed and their performance evaluated primarily by using the energy balance of the first law.

Although the application of the second law has not been extensive, there are many reports and books that have promoted the use of the second law, and many of these provide convincing arguments in its favor (Gardner and Smith 1957; Bosnjakovic 1960; Trepp 1961). However, the engineering literature shows that the second law of thermodynamics has not made the full transition from a textbook phenomenon to an energy-system design tool that is important now for energy conservation.

One reason for the lack of acceptance of the second law by engineers has been the complexity of the concept and the difficulty of defining entropy. Adkins (1968) states, "One of my most difficult problems has been how to present the second law." Callen (1960), in addition to presenting the classical approach to the second law, describes an empirical approach in which the existence and the properties of entropy are stated by a set of axioms:

The postulatory formulation of thermodynamics features states rather than processes as fundamental constructs. Statements about Carnot cycles and about the impossibility of perpetual motion do not appear in the postulates.

This concept of considering entropy at a given steady-state condition rather than as a change in a process is also one of the features of the exergy method.

The concept of available work as initially introduced by Gibbs and Helmholtz for chemical systems is also basic to the exergy method. Available work was developed further by Keenan (1951), and Berg (1974) recently showed the benefits of considering available work in energy conservation.

The exergy method is also related directly to the irreversible production of entropy in systems as so well described by Tolman and Fine (1948) and used recently by Bejan (1975) for analysis of cryogenic equipment and heat exchangers. The concept of minimizing the irreversible production of entropy is inherent in the idea of reducing exergy loss in systems to improve efficiency and conserve energy resources.

Another probable reason for the reluctance of engineers to accept the second law in engineering design is the lack of specific data that show the benefit of using the second law for design purposes. Most textbooks show the effect of second-law considerations in processes, but examples of systems or cycles are usually done with the first-law heat balance only. The more extensive use of the second law through exergy analyses in the Soviet Union and Europe was probably based on the significant results shown by cryogenic and power-system analyses (Brodyanskii, Appendixes B-1, B-2; Bosnjakovic 1960).

2-1 INFLUENCE OF CRYOGENICS

Although there were several single examples of the use of exergy and the second law to energy-related systems of different types, the application of exergy to the field of cryogenics appeared to have the greatest influence in its continued use. The reason for this interest in the low-temperature region is the low Carnot efficiency and the high relative work required to perform refrigeration at low temperatures as shown in Figure 2-1.

An early application of the second law to cryogenics was made by Kapitsa (1939), who proposed that the definition of system efficiency be based on second-law performance. The results of Kapitsa's second-law analysis of cryogenic systems led to his invention of the expansion turbine that efficiently produced low temperatures for air-liquefaction plants.

The analysis of helium-liquefaction systems led Tolman and Fine (1948) to consider the irreversible production of entropy and the second law of thermodynamics. They developed a general energy equation for the second law that is applicable to many other systems and processes. This general equation for net work performed on the atmosphere by a system is

$$\text{Work} = \sum (E - T_0 \, \Delta s) + \sum \frac{T_n - T_0}{T_n} Q_n - T_0 \Delta s_{\text{irrev}} \qquad (2\text{-}1)$$

where the first term on the right involves the energy and entropy transferred between the system and its surroundings, the second term is the heat transferred

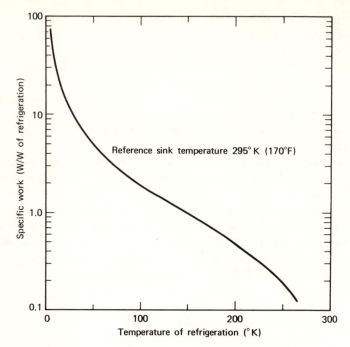

Figure 2-1. Carnot cycle work as a function of temperature level.

to the surroundings at temperature T_n, and the third term is the irreversible production of entropy within the system. This equation also represents the exergy of a given point in a system.

The analysis of other refrigeration systems using the second law includes the work of Ball (1954) and Gardner and Smith (1957). Trepp (1961) performed an exergy analysis in his report on cryogenic refrigeration systems.

The work of Brodyanskii in 1963 involved the direct application of exergy to air-liquefaction systems. His two reports are included in this book as Appendixes B-1 and B-2. Brodyanskii made a detailed analysis of the open-cycle Linde air-liquefaction system and showed the results of an exergy analysis to significantly differ from those for the conventional heat-balance energy analysis. The exergy efficiency was considerably lower than the heat-balance efficiency, and the locations of the high losses in the system were different in the exergy analysis. The results showed that the conventional heat balance gave misleading information in regard to where improvements can be made to raise the system efficiency.

Kun and Ranov (1965) discussed the efficiencies of low-temperature expansion machines, including the influence of irreversible losses, and Meltser (Appendix B-4) considered the exergy method in his evaluation of cooling machines.

Cryogenic insulation performance using exergy was the subject of a paper by Martynovskii (Appendix B-7), and Bejan (1974) analyzed the heat loss through structural supports of cryogenic apparatus using the principle of minimum entropy production.

Because of the need to consider the second law of thermodynamics in cryogenic engineering, the application of exergy, entropy-production analysis, and the second law was extended over the years, and the benefit of using these methods has been demonstrated. The application has not been as well accepted in other areas of our industry. In many cases the use of the second law provided additional information on the system performance that resulted in developing new techniques to improve system performance — Kapitsa's turboexpander and Brodyanskii's secondary coolant loop for liquefaction systems.

Although the second law of thermodynamics has provided significant benefit to cryogenic system design in the past, its continued use is dictated by the advent of new requirements. Haskin and Dexter (1979) point out the low efficiencies and high power requirements being encountered in cryogenic refrigerators for space satellites. A large potential refrigeration requirement exists for liquefaction of natural gas and cooling of long-distance electric power lines and superconductive electrical machinery. The recovery of the exergy in liquefied natural gas during gasification (Oshima et al. 1978) may become an important energy-conservation process. All these future systems will benefit from the use of exergy or the second-law analysis in establishing the most effective system configuration from an energy-conservation or minimum life-cycle cost basis.

2-2 EXERGY DEVELOPMENT

The exergy method of analysis has been developed and used in the Soviet Union and Europe, primarily Germany and Poland. The earliest use of the word *exergy* has been attributed to Rant in 1956 by Bosjnakovic (1960), Trepp (1961), and Baehr (1962). Prior to that time, and in some cases since that time, exergy has been generally referred to as *work capacity* or *available work*. The term *exergy* has been well accepted and used to provide an integrated development of the second-law principles directed primarily to the application in real industrial systems.

The use of exergy in systems analysis appeared in the early 1960s in several books and technical reports. Bosnjakovic (1960) gave an exergy analysis of a power plant in his book on technical thermodynamics using the enthalpy–entropy diagram to present his results. He also made reference to the benefit of using exergy analysis in other energy-related systems. Trepp (1961) examined exergy losses in low-temperature refrigeration machines in a paper presented at an early Cryogenic Engineering Conference at Ann Arbor, Michigan. Trepp used the exergy losses to develop the optimum refrigeration cycle design. Baehr (1962) provided an extensive discussion on exergy in his thermodynamics book with several examples of exergy calculations for power plants. Baehr presented his analytical results by comparing flowcharts of exergy and energy calculated by the first-law heat balance. His results clearly showed the significant difference obtained by these two methods of analysis.

As noted before, Brodyanskii (Appendixes B-1, B-2) wrote two papers on air-liquefaction systems, including a detailed analysis of the Linde system. He presented his results in a table that not only showed the change in exergy in a process or component, but also separated the total exergy-change, useful work, and the different types of irreversible loss. This method of presenting exergy analysis results is very useful for the practicing engineer in showing the location, type, and magnitude of the losses in a system. Although these were Brodyanskii's first reports on exergy, they were based on work presented previously (Ishkin and Brodyanskii 1952, 1958 referenced in Appendix B-1).

In his thesis Evans (1969) stated:

In 1963, Tribus suggested to the writer that Rant's term "exergy" would be a more appropriate name for the property $\mathscr{A}$ than "available energy." The term "exergy" seemed more appropriate since the property $\mathscr{A}$ need not be viewed as a *kind* of energy at all (as such terms as "available energy" and "availability" would imply) but instead may be viewed as a measure of the departure of a system from equilibrium.

This concept of degree of departure from equilibrium (the reference "dead" state) as a measure of exergy or available work is a basic tenet of the exergy method that simplifies the physical logic of the method.

Considerable work was done by Tribus and his co-workers at Dartmouth College using the exergy approach, including several papers on the thermo-economic design of saline water systems (El-Sayed and Aplenc 1964). This group later reverted to the term *essergy* when a more general expression of potential work was developed.

The recent work on the exergy method of analysis has been done by Soviet engineers, and several of their reports have been translated and included as appendixes to this book. These reports discuss the application of exergy to insulation systems, cryogenic systems, and combined refrigeration and heating systems.

2-3   RELATED DEVELOPMENTS

The exergy method of analysis is a particular approach to application of the second law of thermodynamics to engineering systems. The developments in the use of the second law in areas other than exergy can be of benefit in understanding the second law and why the exergy method simplifies its application.

Evolution of the developments relating to the second law is mostly contained in thermodynamic textbooks that have been written since Carnot, Clausius, and Kelvin. Early references to the second law generally reviewed the work of Carnot and Clausius, expanding on the theories with their own ideas. Many of these new ideas were developed in an attempt to rigidly define entropy and prove the principles involving entropy. As a result, much of the textbook information in classical thermodynamics is only indirectly useful to the application of the second law and actually may have hindered its application by introducing unnecessary complexity. The development of these exergy-related concepts are reviewed here for the general background that it provides.

The early work in thermodynamics was directed to the analysis of ideal reversible cycles, and a strong background in the first law of thermodynamics was developed during this period. In the 1930s the analysis of actual thermodynamic system performance was accelerated and the aspect of degradation of entropy and irreversibility examined. It is at this point where much of the classical thermodynamics deviates from a simple exergy approach.

Lewis and Randall (1923; revised 1961) discuss the degradation of entropy in actual thermodynamic systems and point out that loss of energy is not involved but that the loss in the availability of the energy to perform external work does occur as a result of entropy degradation. This is similar to the basic concept of exergy, namely, that energy cannot be lost but only inefficiently used. When energy is inefficiently used, some of the work available is dissipated to overcome the loss, and there is less energy available to perform useful work.

This simple explanation of the effect of lost work as a result of inefficiencies has been overshadowed in classical thermodynamic textbooks by relating the lost work to the irreversibility of a process and then bringing in the concept of the ideal process and system. In developing these concepts, some complicated explanations were devised. Lewis and Randall (1923) used a spring–reservoir system, whereas others used reversible Carnot engines (e.g. Allis and Herlin 1952) and elevators with infinitely small weights (Perry 1963). In addition, the principle of entropy change has been examined from a statistical standpoint. Although the statistical analysis provides further insight into the physical nature of entropy, it applies mostly to extreme cases and is not necessary or desirable to consider when applying the second law to the analysis of energy-related systems. Since the second law has not been widely accepted by engineers, it is apparent that these explanations have not been effective.

As noted before, Callen (1960) used an axiomatic approach to the second law to simplify the application. Others have viewed the second-law analysis from the "available-work" or "lost-work" standpoint, but these presentations have generally been delegated to a secondary role in thermodynamic textbooks.

Keenan (1951) has written specifically about available work, but the primary purpose of the report was to prove the principles involved not to show the application. Others who have included the available work or lost-work concepts in their thermodynamic textbooks are Sonntag and Van Wylen (1971), Howerton (1962), and Faires (1938). The term "available work" is synonymous with exergy, and the effort performed in this area can be directly related to exergy analysis. The term "available work" does have some disadvantages since "available energy" is used relative to our energy resources, and the term "availability" is basic to the reliability and operating capabilities of industrial plants.

In the power-plant field theoretical regenerative steam cycles were proposed as a standard (Selvey and Knowlton 1944), and irreversibilities in the steam cycles were analyzed (Hansen 1945). The evaluation of steam power plant losses based on entropy increases (Keller 1950) provided results similar to that obtained by the exergy method. The comparison of the entropy balance diagram with the heat balance diagram for a plant is similar to the exergy and heat-balance comparisons given by Baehr (1962). However, the entropy-balance diagram does not provide the direct magnitude comparison that the exergy diagram does, although it provides the correct distribution. Kalafati (1963) used the entropy-production principle in his analysis of nuclear power-plant performance, and Gidaspow (1978) recently used this principle in developing two-phase flow equations.

The previous references are only a small part of the literature on the second law of thermodynamics. This field has been studied by physicists and engineers since Carnot and Kelvin but is still an open and controversial subject. Papers and textbook coverage of the application of the second law of thermodynamics has been sporadic, but it does reflect the strong belief of the authors that the application of the second law of thermodynamics is an important aspect of engineering system design. With the increased emphasis on energy conservation that is now in vogue, the pioneering efforts of these authors should be implemented and extended to the point where second-law analyses will be a common method for engineering system design and performance analysis. The simplicity of the exergy method of analysis should aid in this effort.

CHAPTER 3

The Physical Basis of Exergy

The exergy method of analysis is a technique of using the second law of thermo-dynamics in actual systems analysis. It has been developed to avoid much of the complexity and confusion that exists in the classical approach to the second law of thermodynamics. It is only necessary to realize that work can be performed only under conditions that are not at rest in the surrounding environment. More work can be performed when the conditions are farther from equilibrium with the surrounding environment. Work is performed as the state of conditions returns to equilibrium with the surroundings, as all matter will eventually do. The exergy method is concerned with how well we use the available work that is generated from our energy resources. It answers the questions of where, why, and how much of this available work is lost in our systems. This is the information required to design and operate more efficient systems from the energy-conservation viewpoint.

The application of the exergy method requires the extensive use of entropy. This poses no problem since entropy is considered as a property similar to enthalpy and is listed in tables for many materials. In those cases where tabular values of entropy are not available, entropy can be calculated from

$$ds = \frac{dQ}{T} \tag{3-1}$$

Since $dQ = (dU + w)$, the entropy equivalence for any form of work or energy can be calculated.

Complete understanding of the concepts of reversibility, irreversibility, perpetual-motion machines, thermodynamic probability, and so on is not required for the practical application of the exergy method. These concepts are adequately covered in many of the thermodynamic textbooks listed in the bibliography, and their discussion here would only hinder the engineering application of the exergy method. It is the author's hope that through the application of exergy analysis, the actual benefits of using the second law of thermodynamics will be realized and that the interest in this fascinating field of science and engineering will accelerate.

The fundamentals of the second law of thermodynamics are discussed in this chapter to the extent that they relate to the application of the exergy method of analysis to engineering systems.

3-1   AVAILABLE WORK

The exergy method of analysis is based on evaluating the work that is available at various points in a system. From an analysis of the available work throughout a system, the quantity and location of lost work and useful work can be determined. This is the information required to make a complete exergy analysis of a system and to locate inefficient processes, equipment, or operating procedures. There are two approaches to evaluating the available work in a system, one for designing new systems and one for evaluating existing systems.

Available work is calculated on the basis of a final heat-sink reference that is generally taken as the surrounding environment. The available work at any given point in a system is

$$\text{Available work} = (h - h_0) - T_0(s - s_0) \tag{3-2}$$

For a given process the change in available work from point 1 to point 2 is

$$\text{Available work change} = (h_2 - h_1) - T_0(s_2 - s_1) \tag{3-3}$$

If no useful work is done in the process, this change is a loss in available work.

In the design of a new system involving the generation or use of energy, the exergy method will provide the information to better select the component designs and operating procedure that will be the most effective on any basis of selection, such as plant and operating costs, energy conservation, fuel versatility, and pollution. The design of a new system is performed by evaluating the losses in work associated with candidate processes, equipment designs, and operating procedures that can be used in the system to provide the final product. This analysis must be done on an integral system basis since the loss in work as a result of irreversible inefficiency in processes and hardware components will vary because of the boundary conditions imposed on the process or component. It should be noted that after a process or component is selected from a minimum work-loss analysis based on system operating assumptions, an iteration with varying boundary performance conditions should be made to assure that the overall system conditions meet the requirements best. The effect of operating conditions on the system efficiency is much stronger in the lost-work analysis than it is for the heat-balance analysis. For example, throttling of steam from 1000 psia and 1000 °F to 800 psia in an environment of 70 °F will result in no change in the enthalpy but a loss in available work:

$$\Delta h = 0$$

$$\text{Loss in available work} = T_0 (s_2 - s_1) = 530(1.6836 - 1.6603)$$

$$= 12.33 \text{ Btu/lb steam flow}$$

If this same process is performed with the steam at 600 °F instead of 1000 °F, there is still no change in enthalpy but the loss in available work will be

$$\text{Loss in available work} = T_0 (s_2 - s_1) = 530(1.4657 - 1.4455)$$

$$= 10.69 \text{ Btu/lb steam flow}$$

Therefore, in a heat-balance analysis it will make no difference if the throttled steam is at 1000 °F or 600 °F, but the exergy analysis shows that the loss in available work in the 1000 °F steam is 15% greater than the loss in the 600 °F steam.

The maximum work available is in the initial energy source of the system. For a gas-fired steam generator, this would be the heat content of the gas. If all the available work in the gas when burned were extracted as useful work, the system would be 100% efficient. However, in the process of combustion and heat transfer to the steam, up to 50% of the available work is lost as a result of irreversible processes. At the system interfaces with the environment, available work is lost in the condenser cooling water, heat in the flue gas, and heat leakage from the warm components such as furnace walls.

The system design is developed by iteratively varying the process and equipment performance characteristics until the best system is determined. One parameter used in system selection is the plant efficiency, which is usually based on a heat-balance analysis. A true measure of the plant efficiency is based on the available-work principle:

$$\eta = \frac{\text{Useful work}}{\text{Available work}} = \frac{\text{Available work} - \text{lost work}}{\text{Available work}}$$

The minimization of lost work in the system will provide the most efficient system. The exergy method of analysis is based on this principle.

In an exergy analysis of an existing system, the primary task is to determine where losses in available work exist in the system, what the magnitude of these losses are, and which of the losses, when corrected, can most effectively improve the system efficiency. Many situations exist where lost work is associated with old and worn equipment. Replacement of the equipment would improve system efficiency but can be a costly procedure. At the other extreme there are situations where simple and inexpensive modifications such as adding thermal insulation or sealing gas leaks will improve the system efficiency. It is most important, however, to have a true measure of the loss in work that exists in the system and the locations and magnitudes of the losses in available work. From measurements made throughout an existing system, an energy balance can be made that forms the basis for determining the available work throughout the system. Examination of the changes in available work along with information on useful work in the system will show the locations and magnitudes of losses in the system. The selection of losses in the system to be corrected must be based on good engineering—economic judgement. However, the use of the available-work principle in the exergy method will provide a significant improvement in the data used for making the judgments since it considers the quality of the inefficient energy use as well as the quantity.

3-2   THE REFERENCE SYSTEM

A basic principle of the exergy method of analysis is that the calculation of the exergy (or available work) is related to a common reference condition at which the system working matter loses the capability to do further work. This reference condition is generally the surrounding environment of the system since the final disposition of the initially available energy will be through external cooling systems or ejection of energy-bearing matter into the environment.

The use of the surrounding environment as a zero-energy reference offers many advantages for engineering design and analysis of energy-related systems. For example, the discharges of waste heat from different industrial plants are compared on the basis of the number of Btus. Energy consumption and requirements are also compared on the basis of the number of Btus. These comparisons are developed from heat-balance data and show only the quantity of energy. The fact that one plant requires its Btus to be high-quality energy at $1000\,°F$ for metal processing whereas another plant only needs its Btus at $300\,°F$ to heat a plating bath is not shown by these comparisons. A very important aspect of energy conservation and effective energy use is missing from these comparisons, which can be satisfied by using the exergy method referenced to the surrounding environment.

The use of a common reference base, such as the surrounding environment as the ultimate heat sink and using work as the basis of evaluation, provide the commonality required to make valid comparisons of energy consumption and requirements not only between the same type of industrial plants, but between different types of plant. For example, power-generating plants can be compared to industrial refrigeration plants with respect to efficiency of using energy resources.

The use of the surrounding environment as a reference for exergy analysis is most practical since it is the ultimate sink for all energy-related processes. However, there are situations where some other reference would be more suitable in a particular analysis. For example, if an existing process using direct-fired fossil fuel is to be converted to using the exhaust heat from a power-generating system, the selection and design of the power-generating system would be based on a reference related to the input requirements of the industrial process. Although sea level would be the ultimate sink for fluid-flow systems, a local reference plane is more practical to take for analysis and is completely suitable since it is the difference in gravity levels that relates to work performed or required.

Under some conditions the surrounding environment may not act as an infinite heat sink, and the reference may vary with the influence of the energy system being evaluated. For example, when dissipating waste heat to the atmosphere through a dry-heat exchanger, the atmospheric air in the heat exchanger is heated up. This can be corrected for in the exergy analysis by either correcting the reference condition to account for this heating effect or by adding another lost-work step in the system and maintaining the original atmospheric condition as the final reference. Brodyanskii and Kalinin discuss this in Appendix B-3.

The most important point concerning a reference system for exergy analysis is that it is a tool that can simplify and extend the application of exergy analysis as well as provide the basis for the second-law analysis. Small variations in the environmental conditions are important to consider only when direct comparisons of different energy systems are involved. In this case the system exergy data should be corrected so that the comparisons can be made with a common reference condition.

In most cases with common gases such as air and helium, the use of a reference of the gas at the temperature and pressure of the surrounding environment will result in all values of exergy being positive as exergy should be relative to the "dead" state. However, for some fluids such as the Freon refrigerants that do not have a common temperature and pressure conditions with the environment, negative exergy values can be calculated for a common reference state. In a closed cycle this condition is satisfactory since exergy differences are of interest in analyzing system performance. Negative values of exergy should be considered positive when the fluid is taken as a single state referenced to the surrounding environment. For some fluids it is generally convenient to assume the reference state as a liquid at the temperature of the surrounding since the compressibility of the liquid is low and little exergy change is involved with liquid pressure change.

When a fluid mass is injected into the environment, there is mixing with the surrounding atmosphere. This mixing does involve exergy losses resulting from the irreversibilities of mixing. However, such losses are generally neglected, at least at present, since recovery of the mixing losses is not practical. The exergy involving the momentum and chemical potential of a mass exhausted into the atmosphere is usually negligible relative to the system exergy changes. In special cases these exergy losses can be calculated and considered in the exergy balance of the system.

3-3   GENERAL EXERGY EQUATION

The basic procedure for conducting an exergy analysis of a system is to determine the value of the exergy at steady-state points in a system and then the cause of the exergy change for the processes that occur between these stations. The calculation of the exergy value is performed using the characteristics of the working medium established from a first-law energy balance.

Since exergy is defined as the work available at a steady-state point in the system relative to a reference "dead" state (the surrounding environment), a general exergy equation can be made up as a sum of all the exergies that contribute to the available work at that point. The exergy components that make up the total exergy will differ from system to system, and a general equation would have a large number of terms that would be cumbersome. The approach taken here is to tabulate the common contributors to the work in most systems so that the exergy equation can be formulated from these terms as required for any system.

The general form of the exergy equation is developed using Figure 3-1, which shows $T-s$ diagrams of a heat engine. If a given amount of heat Q_H is introduced into the system at temperature T_H, the equivalent amount of work could be extracted only if this thermal energy were dissipated through an ideal work-recovery device (engine with no losses) to a temperature of absolute zero T_A (Figure 3-1a). All the input heat could be converted to work, and the exergy at T_H would be

$$Ex_H = Q_H \tag{3-4}$$

When the heat from the work recovery engine is rejected to a finite-temperature heat sink T_0 (such as the surrounding environment on earth), only that exergy between T_H and T_0 is recoverable and that part below T_0 is lost (Figure 3-1b). Therefore, the available work or exergy at T_H is now

$$Ex_H = Q_H - T_0 \, \Delta s \tag{3-5}$$

where $T_0 \, \Delta s$ is the nonrecoverable part of the initial ideal exergy. The term $T_0 \, \Delta s$ in eq. (3-5) represents Q_0, the heat rejected reversibility to the surrounding environment (no finite-temperature difference). If we consider that the enthalpy

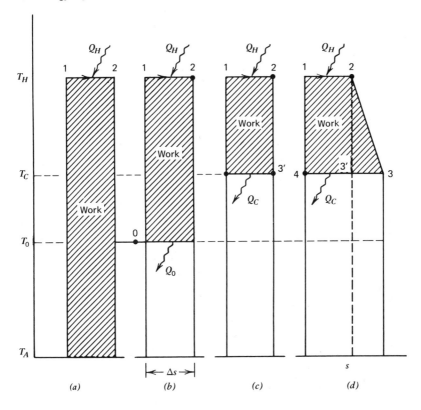

Figure 3-1. Closed-cycle $T-s$ diagram for a heat engine.

represents Q for a unit mass (steady-state, steady-flow conditions assumed), the exergy at T_H with heat sink at T_0 is

$$Ex_H = h_H - h_0 \qquad (3\text{-}6)$$

However, the general exergy equation must consider the losses that occur in an actual system. Heat cannot be rejected from an engine to the surrounding environment without a temperature difference to cause the heat flow. This condition is shown by Figure 3-1c, where heat is rejected to the surrounding environment T_0 at a temperature of T_C. The portion of the initial exergy at T_H that is lost through this heat transfer is

$$Ex_{\text{loss}} = Q_C \left(1 - \frac{T_0}{T_C}\right) \qquad (3\text{-}7)$$

Expanding eq. (3-7) gives

$$Ex_{loss} = Q_C - T_0 \frac{Q_C}{T_C}$$ (3-8)

but Q_C/T_C is equal to the entropy s. The exergy loss from rejecting heat to the surroundings at T_C is the heat Q_C minus any irreversible production of entropy generated in the system. The distinction is made between entropy production within a system and outside a system for compatibility with the reversibility concept that is generally accepted in thermodynamics textbooks. The system shown in Figure 3-1c is internally reversible between temperatures T_H and T_C, but it is externally irreversible since it rejects its heat to the surroundings with a finite-temperature difference, $T_C - T_0$.

The irreversible production of entropy within a system because of friction, heat transfer with a finite-temperature drop, or other inefficiencies results in loss of exergy in the system. Figure 3-1d shows a typical irreversible loss inside the system that causes production of entropy from s_2 to s_3 without an equivalent production of useful work. This exergy loss is represented by the area under the T_0 line between s_3 and s_2.

$$Ex_{loss} = T_0(s_3 - s_2)$$

or, in general,

$$Ex_{loss, internal} = T_0 \Delta s_{irreversible}$$ (3-9)

The preceding explanation of exergy used a closed-cycle system to describe the effect on the system work by various losses that are inherent in the system or that result from inefficient design or operation. The use of the closed cycle has allowed depiction of system work by areas on the diagrams. The exergy value we are interested in for an exergy analysis is of a steady-state point in the system relative to the surrounding environment. Point 2 in Figure 3-1b has the highest value of enthalpy and entropy in the cycle following the isothermal heat-addition process 1−2. Point 0 is the reference condition and has the enthalpy and entropy values of the working fluid in the system as if it were at rest under the surrounding ambient conditions.

The terms that would make up a general exergy equation for a steady-state point in a system are listed in Table 3-1 for the more common types of energy-related system. Diagrams depicting the exergy relationships for the terms are shown in Figure 3-2.

3-4 ENTHALPY AND EXERGY

In present engineering practice, systems involving heat or refrigeration use the enthalpy of the working fluid in an energy balance to evaluate the system

Table 3-1 Terms of General Exergy Equation

Type of Process	Driving Potential	Flow Charge	Entropy Change	Exergy-equation Term
Heat flow	Temperature difference	Heat	dQ/T	$T_0\,dQ/T$ or $T_0 c_p\,dT/T$
Electrical	Voltage difference	Current, I	$I^2 R/T$	$T_0 I^2 R/T$
Flow momentum	Velocity	Mass flow, G	$(c_i + c_2 \cdots)\dfrac{V_2^2 - V_1^2}{2gJT}$	$T_0 \sum c_i (V_1^2 - V_0^2)/2gJT$
Gravity	Altitude difference, z	Mass, m	$m(z_2 - z_1)(g/g_c)/J$	$T_0 mg(z_1 - z_0)/TJg_c$
Thermoelectric	Electromotive force, voltage difference	Heat and current	$(dQ/T) + (I^2 R/T)$	$T_0[(dQ/T) + (I^2 R/T)]$
Friction	Pressure drop, Δp	Mass, m	$(P_2 - P_1)/J$	$(P_1 - P_0)T_0/JT$
Melting, freezing	Enthalpy difference	Phase change	$\Delta h/T$	$T_0\,\Delta h/T$
Chemical reaction	Chemical potential	Heat, Q	$N_i(\mu_{i,1} - \mu_{i,0})$	$T_0 N_i(\mu_{i,1} - \mu_{i,0})/T$
Radiation heat transfer	Fourth-power temperature difference	Heat, Q	$\epsilon A F \sigma (T_1^4 - T_0^4)$	$\epsilon A F \sigma (3T_1^4 - T_0^4 - 4T_0 T_1^3)$
Mixing (isothermal)	Partial pressure	Mass, m	$R_i \ln(P_{i,2}/P_1)$	$T_0 \sum R_i \ln(P_{i,2}/P_1)/T_i$

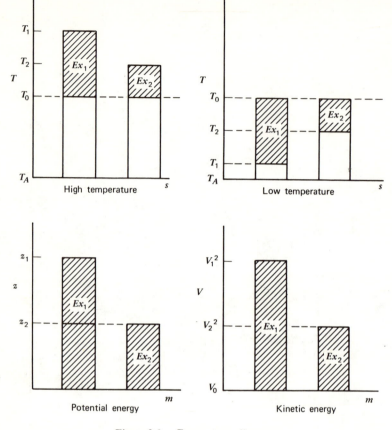

Figure 3-2. Exergy-term diagrams.

performance characteristics. Enthalpy is a function of the state of a substance and is defined as

$$h = I + PV$$

where I is the internal energy of the substance. A change in enthalpy is then

$$dh = dQ + V\,dp$$

where dQ is the change in heat energy. Under a constant-pressure condition

$$dh = dQ = c_p\,dT$$

where c_p is the specific heat of the substance at constant pressure. Specific heat is defined as the number of heat units (Btu or calorie) required to raise the temperature of water one degree per unit mass. Engineering units are Btu/lb °R or cal/gm °K. Therefore, specific heat is the heat content of a mass per unit of temperature gradient, and enthalpy is the total heat content of the mass for a given temperature. The variation of c_p for air over the temperature range 100– 1000 °F is only about 10%. A 50 °F temperature change at these temperature levels would result in the following enthalpy values

$$\Delta h_{565°R} = h_{585°R} - h_{535°R}$$

$$= 139.9 - 127.9 = 12 \text{ Btu/lb}$$

$$\Delta h_{1460°R} = h_{1485°R} - h_{1435°R}$$

$$13.1 \text{ Btu/lb}$$

These calculations indicate that a 50 °F temperature drop in air will transfer only 10% more heat at 1000 °F than it does at 100 °F. The work available in the 50 °F temperature drop at 1000 °F is

$$\Delta Ex = (365.21 - 352.06) - 530(0.85151 - 0.84250)$$

$$= 8.37 \text{ Btu/lb}$$

and for the 50 °F temperature drop at 100 °F,

$$\Delta Ex = (139.86 - 127.86) - 530(0.61999 - 0.59855)$$

$$\Delta Ex = 0.64 \text{ Btu/lb}$$

The transfer of heat from the airstream with a 50 °F temperature drop results in thirteen times greater exergy or available energy for work at 1000 °F level than at the 100 °F level. Figure 1-25 illustrates the effect of temperature on enthalpy and exergy.

The primary use for heat-generated energy is to provide work, and in this respect the 50 °F temperature drop at 1000 °F level is much more valuable to us than the 50 °F temperature drop at 100 °F. The heat content of a gas as represented by the enthalpy value does not show the true work capability of the gas. This is because the enthalpy shows the quantity, but not the quality, of the heat in the gas. The quality of the heat and the work available in a gas is repre- sented by the exergy value of the gas. Since the primary use for our energy sources is to perform work, the quality of the available energy in our resources must be considered if we are to make maximum use of the available work in our energy resources.

3-5   EXERGY LOSS, REVERSIBILITY, AND IRREVERSIBILITY

The most efficient use of the exergy supplied to a system is achieved when the exergy losses in the system are minimum. A primary use of the exergy analysis is to show the location, type, and magnitude of the exergy losses in a system so that the system efficiency can be most effectively increased by reducing these losses. These exergy losses are generated throughout the system by the irreversible production of entropy caused by the nonideal performance inherent in all real systems and components.

Classical thermodynamics textbooks dwell extensively in the idea of reversibility and irreversibility of processes and cycles. Although exergy losses are related to the irreversible characteristics of a process or system, the exergy method can be applied and understood without extensive involvement of irreversibility. Under some conditions the use of reversibility and irreversibility in the evaluation of process efficiency is controversial and confusing. For example, the turbine expansion process in an ideal reversible Carnot cycle (Figure 3-3) is an isentropic process (1–2) between two pressure levels p_1 and

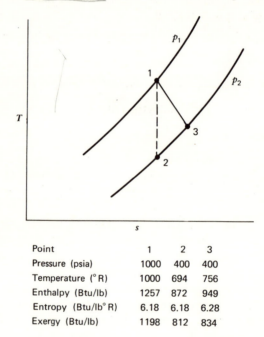

Point	1	2	3
Pressure (psia)	1000	400	400
Temperature (°R)	1000	694	756
Enthalpy (Btu/lb)	1257	872	949
Entropy (Btu/lb°R)	6.18	6.18	6.28
Exergy (Btu/lb)	1198	812	834

Figure 3-3.   Reversible and irreversible processes.

p_2 with no irreversibilities. In an actual turbine-expansion process between the same parameters, losses exist that make the process irreversible (1–3). The steam or gas condition leaving the turbine is changed to achieve the final pressure. Now the question arises as to how the efficiency of this process should be defined. It is often defined as the actual efficiency (work out divided by the energy in) relative to the ideal efficiency for the isentropic process. Since the ideal case cannot occur and the conditions at the outlet of the turbine are not an actual set of system operating conditions, this definition has little meaning to the system engineer. In the actual turbine operation the total work into the turbine minus that leaving the turbine is dissipated in useful shaft work, with the remainder lost through irreversibilities such as friction, turbulence, bearing and seal wear, and leakage. A more practical efficiency is defined as the actual shaft output divided by the energy into the turbine. This efficiency is a direct measure of the relative losses in the component and can be related to the other losses in the system. All ideal processes are reversible, and the introduction of inherent but undesirable losses results in irreversible characteristics in the process. It is the goal of good engineering to minimize these irreversibilities in those places where they contribute significant exergy loss to the system.

The calculation of exergy at discreet points in a system will indicate the total exergy change, reversible and irreversible, in the process and in the equipment between any two points. These exergy values are explicit values relative to the reference condition that is usually the surrounding environment. It is thus necessary to separate this total exergy into the actual work contributions from the work-loss contributions that are attributable to irreversibilities. This aspect of the exergy-analysis method requires a knowledge of the degree of reversibility of the total process and additional calculations to identify the individual irreversibilities or exergy losses. This phase of the analysis, however, can be controlled in its degree of sophistication as a function of the requirements of the analysis. For example, in a preliminary analysis of a closed-cycle Brayton power system the total turbine exergy loss can easily be calculated using an empirical isentropic efficiency derived from test data of similar turbines. The exergy loss obtained in this manner would be approximate but suitable for preliminary system analysis. When a full design and performance analysis is required, the individual exergy losses associated with the mechanical friction, mass, and heat leakage and the fluid-flow turbulence and friction would be analyzed as extensively as possible to assure that the specific locations and causes of these exergy losses are determined as realistically as possible. A tabulation of these irreversible losses in a system will indicate the relative magnitude of these exergy losses.

In addition to the magnitude of the exergy losses, the causes for these losses must be considered since some types of exergy loss will require complex and expensive designs or modifications to reduce the losses whereas other types of loss may be reduced by simply adding thermal insulation or sealing up flow leakages.

To illustrate further the location and types of exergy losses in energy-related systems, the typical losses for a steam power plant are shown in Figure 3-4. Table 3-2 lists these losses, describes the causes of the losses, and discusses the methods and complexities of reducing the losses. The relative magnitudes of the

58

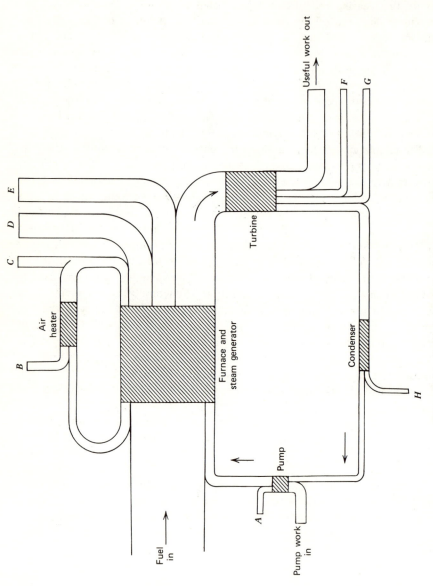

Figure 3-4. Typical exergy losses in a power plant.

Table 3-2 Exergy Losses in a Power Plant

Exergy Loss	Type of Loss	Cause of Loss	Method of Reducing Loss	Significance of Loss in System
A	Pump inefficiency	Friction, turbulence in fluid, bearing and seal friction	Improve component design	Pump loss insignificant in this system
B	Air-heater irreversibility	Heat transfer through temperature drop and pressure drop	Improve heat transfer and reduce pressure drop through design research	Air-heater exergy loss small relative to exergy recovered
C	'Flue gas to atmosphere	Rejection of heat to atmosphere	Reduce rejection temperature or recover heat for industrial process, space heating, or drying	Large number of Btus lost this way but quality low; not a major loss in energy-intensive systems
D	Combustion	Irreversible energy transfer from chemical to heat	Inherent in system; cannot be reduced with present technology	Important exergy loss in most energy-intensive systems
E	Heat transfer, combustion gas to steam	Heat transfer through temperature gradient	Raise steam temperature, use topping cycle, improve heat transfer	A major source of exergy loss in energy-intensive systems
F	Mechanical friction	Bearing and seal friction	Improve design of component and improve lubrication	Generally small relative to other losses
G	Hydraulic	Fluid-flow turbulence, windage	Improve design of fluid-flow passages	Significant exergy loss in most systems
H	Heat rejection to condenser cooling water	Rejection of heat to surrounding environment	Reduce rejection temperature; recover heat for space heating, industrial process, or drying	Large Btu rejection but low in quality, so exergy loss is not significant relative to other exergy losses

59

losses can be shown in Figure 3-4 by the width of the loss band for any specific system analyzed.

A typical refrigeration system is examined in this same way with Figure 3-5, showing the distribution of the exergy losses and Table 3-3, describing the losses and discussing the methods and the complexities of reducing the losses.

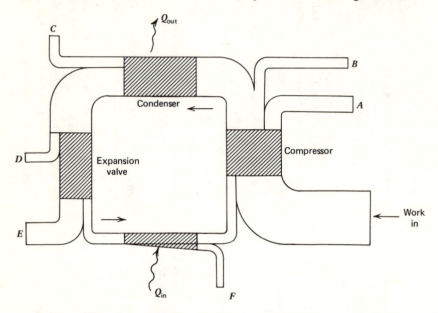

Figure 3-5.   Typical exergy losses in a refrigeration system.

All irreversibilities in processes and systems should be given consideration if effective energy-conservation measures are to be conducted. In many cases the irreversible exergy losses are small and do not warrant serious design or cost effort. In other cases these irreversible exergy losses may be relatively large, but the technology is not at hand to make the desired reductions in the loss. For example, reduction of the large irreversible heat-transfer loss between high-temperature combustion gas and steam requires the development of improved high-temperature metals to permit the use of higher steam temperatures, or either a magnetohydrodynamic (MHD) or liquid metal circuit could be used as a topping cycle to recover work. Research–development efforts appear warranted in technology areas where the reduction in large, irreversible losses is limited by existing engineering technology.

The reason for considering all exergy losses in a system during initial stages is that some processes that usually have a low exergy loss may show up to be the most significant loss in another type of system. The throttling process in a conventional home refrigerator system makes only a small contribution to the

Table 3-3 Exergy Losses in a Refrigeration System

Exergy Loss	Type of Loss	Cause of Loss	Method of Reducing Loss	Significance of Loss in System
A	Compression inefficiency	Friction, fluid turbulence	Improve component design	Significant loss in most refrigeration systems
B	Duct pressure drop	Frictional loss in duct and fittings	Improve design and lower flow rates	Usually small loss relative to other losses
C	Heat rejection to heat sink	Rejection of heat to surroundings	Lower rejection temperature to minimum practical; recover heat for other uses	Large number of Btus lost this way but quality low; not a major exergy loss in most systems
D	Piping pressure drop	Friction loss in pipe and fittings	Design improvement or larger pipe	Usually small loss in refrigeration systems
E	Expansion valve loss	Isenthalpic processes with no work recovery	Install work-recovery device such as hydraulic turbine	Significant loss in low-temperature systems
F	Heat transfer in evaporator	Heat transfer through temperature drop and pressure drop	Improve design and reduce temperature gradient and pressure drop	Loss can be significant in low-temperature systems

total system exergy loss, but when the throttling is used at cryogenic temperatures for liquefying air, the exergy contribution is a major factor in the total
system exergy loss.

The presence of a high exergy loss in a system may indicate a poor match
between the quality of the energy supplied to a process and the quality required
by the process to perform a given task. This mismatch results in large irreversible
losses. The primary example of this mismatch is the conventional gas-fired hot-
water heater where the high-quality gas flame at about 3000 °F is used to heat
water at less than 150 °F. Although these hot-water heaters are given an
efficiency of around 50% based on 50% of the input Btus being lost up the flue,
the actual exergy efficiency is

$$\eta = \frac{(T_c - T_0)/T_c}{(T_h - T_0)/T_h} = 0.15$$

because of the large irreversible heat transfer between the gas flame and the
water. A better match between the energy supplied and the energy required
would be made if the energy source were available in the temperature range
200–400 °F. Solar panels can supply this energy, although the heat-storage
problem for hot-water heating at night must be realistically and economically
solved. The major benefit of using solar or other low-quality energy sources for
low-quality energy requirements such as home, swimming-pool, and hot-water
heating is that it releases the high-quality energy sources (coal, oil, and gas) for
better matched (high-quality) requirements such as electrical power generation,
transportation, and industrial processes.

3-6   LOSS IN PROCESSES

The analysis of the loss in exergy for energy-related processes has a range of
complexity from the simple case where a single loss is equivalent to the total
exergy change in the process to a process where the total exergy change is
composed of several losses as well as useful work.

In those cases involving a single type of exergy loss, such as frictional loss in
piping and throttling valves and heat-leak loss through duct walls, the total
change in exergy calculated from the properties before and after the flow
process will be the loss in exergy if there is no useful work involved with
surrounding environment. Other processes can involve multiple exergy losses
coupled with work either added to or removed from the process. In these cases
the total exergy change will include the losses as well as the useful work change.
In the exergy analysis these losses can be handled in several ways. For processes
that take place in complex equipment such as a gas turbine, the losses may be
lumped together and separated from the useful work performed. This is the most

common approach, and since the useful work output from a turbine can be directly measured at the shaft during operation, the lumped value of the total loss can be readily determined.

The complexity in exergy analysis comes about when improvements in the efficiency of these processes are desired, since it is necessary to know the magnitude and locations of the individual losses to effectively accomplish this improvement. For example, it would not be practical to expend time and effort to reduce heat leak from a device if the loss from the heat were negligible compared to the other losses in the process. On the other hand, some processes may involve high exergy losses but the technology may not be available to reduce these losses. When these situations are found, the question of initiation research efforts to develop the technology is raised.

The practical approach to exergy analysis of processes is to calculate the total exergy change in the process and separate the external work involved from the total exergy loss. The process exergy loss is then compared with other exergy losses in the system to establish the impact of the specific process exergy loss to the overall system efficiency. If the process loss shows a significant influence on the system efficiency, evaluation of the individual exergy losses in the process is warranted.

The individual losses in a process are determined in the same manner that a process or system is designed. As a matter of fact, the design equations can generally be used to calculate the individual heat losses, mass leakage, friction losses, and other variables, which are then put into the form of exergy losses. Individual losses generally fall into the categories of chemical change, heat transfer, friction, machine inefficiency, and heat rejection. The machine inefficiency may be separated into more discrete loss components such as windage loss or bearing loss if warranted.

As the exergy losses in energy-related process are calculated, tabulated, and compared, the real causes of inefficiency in the process design and operation will be disclosed. With this information at hand, the task of effectively directing efforts in performance improvement and energy conservation is simplified.

3-7 LOSS IN SYSTEMS

The analysis of exergy losses and useful work in a process will indicate the relative magnitude of the losses in the process and, therefore, where improvements in the process can be made to increase the process efficiency. When a process is used in conjunction with other processes in a system, which is the more common case, the relative magnitude of the exergy losses in each process must be considered if effective improvements in the overall system efficiency are to be made. It is in this evaluation where the usefulness of the exergy method is noted. A comparison of energy losses developed from a first-law energy balance will provide only a quantitative comparison of the losses. A comparison of exergy losses will include the quality of the losses as well as the quantity of energy. This is particularly important in systems where initial energy processes involve high-quality energy transfer whereas other processes in the system involve low-quality energy transfer.

When the exergy losses for each process in a system are tabulated, the initial comparison of process loss will show that many of the processes have little impact on the system efficiency relative to the losses in other processes. At this point the overall losses in the high-energy-loss processes are further broken down to individual losses where practical to indicate the type of improvement that can be made and the relative effect that the improvement will have on the overall system efficiency.

The criteria for final system exergy-loss evaluation are determined by the purpose of the analysis. If the most efficient design is desired on the basis of performance, the minimum total system exergy loss will provide the solution. If the analysis is to be a minimum life-cycle cost evaluation, the exergy losses in the system are related to time dependent costs and initial equipment costs to provide overall cost factors that can be minimized through parametric analysis. If energy conservation is the purpose of the analysis, the tabulation showing the magnitude, type, and location of the exergy losses will provide the information needed to complete the study. The initial magnitude of the individual exergy loss and the potential for its reduction through design or operational change, as well as cost effectiveness, environmental impact, and other parameters of the change are all considered in determining where the energy-conservation measures are to be made.

The tabulation of individual exergy losses in energy-related systems can constitute the basis for determining where research—development efforts are warranted to improve performance or effectiveness of energy use. Because the exergy-loss analysis considers the quality of the energy as well as the quantity, a true measure of the worth of the losses is provided.

3-8   LOSS TO SURROUNDINGS

The loss of mass or thermal energy from a system to the surrounding environment is a direct loss of the exergy in the system. When this occurs with high-quality energy, such as leakage of high-pressure mass or high-temperature heat, the impact on system efficiency and performance can be significant. Fortunately, most of the energy that is discarded to the surrounding environment is of low quality and thus involves relatively little loss of work. The reduction of exergy loss to the surrounding environment is one of the most rewarding approaches to improving system efficiency since in most cases the reduction in exergy loss also improves environmental conditions.

The losses of exergy to the surrounding environment fall generally into two categories. The loss of high-quality energy in the form of mass or heat at high or low temperature or pressure can contribute significantly to degrading the efficiency of a plant. Methods of reducing this exergy loss are well known and simple. One of the most common losses of this type is the heat loss from high-temperature furnaces, process heaters, and ovens used in industrial processes. Sealing hot-gas leaks and improving the thermal insulation in the equipment can go a long way in reducing the exergy losses in the plant. The magnitude of the savings from these improvements is usually greatly underestimated when the savings are calculated on a Btu basis from a first-law energy balance. The true value of an improvement made to reduce losses in high-temperature and high-pressure equipment is given by the second-law principle embodied in the exergy method where the high quality of the lost Btus is considered. Major improvements in industrial plants will be realized when processes involving high-quality energy are put on a continuous-flow basis where energy can be recovered and reused as opposed to the batch process, which involves wasteful reheating. Continuous steel making processes are being developed (Tselikov 1976) in which the hot gases are recycled for efficient system operation. The economics of modifying or building new efficient process plants as opposed to maintaining existing concepts must be weighed, but the benefits in energy conservation, operating costs, and environmental impact are becoming stronger in favor of energy-efficient system designs, especially when the quality of the energy involved is considered.

The loss of low-quality energy to the surrounding environment from our commerical and industrial plants involves a large number of Btus. The recovery and use of this low-quality waste energy has taken a prominent position in our society. The potential of recovery and use of waste heat from industrial plants is generally overestimated when the evaluation is based on a first-law energy balance, since the low quality of the energy is not considered. In general, the cost and size of the equipment required to recover a given amount of work from low-quality energy is much greater than for the same amount of work from the same Btus of high-quality energy. However, the use of this low-quality energy can be of significant benefit if it is used in an application where a high-quality energy source is replaced. For example, if waste heat from a power plant is used to heat industrial metal-processing baths replacing gas or oil-fired heaters, there is a significant improvement in energy conservation, primarily because the high-quality energy source can now be used in a more matched application that requires a high-quality energy source, such as power generation.

The loss of exergy from a system to the surrounding environment is considered in the exergy method of systems analysis the same as internal system losses, and they are included in the system efficiency calculation. When the waste energy is recovered and used in another process, the system efficiency should reflect this improvement in use of the initial available work. On the other hand, full credit for the exergy given to the secondary process is not allowable since the energy must be rejected to the surrounding environment from the secondary process at the same exergy level above the surrounding environment. This situation can be handled in two ways in the exergy analysis. In one case, the primary system is analyzed as if the waste energy were rejected directly to the surrounding from the primary system, and the primary system is credited with useful work of the secondary process. The other approach would consider that the secondary process that uses the waste energy is part of the primary system; in other words, the basic system is now a combined-cycle system. From a practical analysis standpoint, the first method would be used except where the secondary process is a significant contributor to the useful work from the initial available energy.

When materials from an energy-related system are exhausted into the ambient atmosphere, there are additional exergy changes involving the mixing of the exhausted constituents with the atmospheric air. It would be difficult to recover this energy, and it is included in the exergy loss that occurs from direct mass and energy dissipation to the surrounding environment.

3-9   EXERGY LOSS IN HEAT EXCHANGERS

Heat exchangers are major components in most energy systems, and they are the source of significant exergy losses. Heat exchangers are generally inefficient from an exergy standpoint because they have been designed in the past on the basis of low first cost that dictates a minimum-sized unit. To achieve the small-sized heat exchanger, the temperature difference between the fluid streams is maximized. However, the larger is the temperature difference in a heat exchanger, the greater will be the exergy loss during heat transfer. The basis of exergy loss in heat exchangers is discussed here relative to the design and operating conditions.

Example 3-1   To show the procedure in developing the exergy loss in heat exchangers, a simple counterflow air–air unit (Figure 3-6) is considered. The temperature of the hot airstream is held constant while the temperature of the cold airstream is varied to show the variation of the exergy loss as a function of the temperature difference between the hot and cold airstreams. It is assumed that 50,000 Btu/hr of heat is transferred to the cold airstream.

Condition A

$$T_{h, in} = 1150\,°F \qquad T_{h, out} = 850\,°F$$

$$T_{c, in} = 650\,°F \qquad T_{c, out} = 950\,°F$$

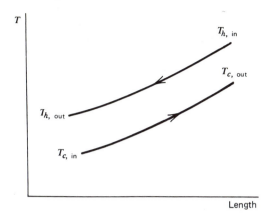

	T (°F)	h (Btu/lb)	s (Btu/lb °R)	Ex (Btu/lb)
Hot-side condition				
$T_{h,in}$	1150	398.4	0.8930	125.1
$T_{h,out}$	850	319.5	0.8188	74.9
Cold-side condition				
A $T_{c,in}$	650	268.5	0.7765	46.3
$T_{c,out}$	950	345.5	0.8379	90.8
B $T_{c,in}$	550	243.5	0.7529	33.8
$T_{c,out}$	850	319.5	0.8188	74.9
C $T_{c,in}$	450	218.7	0.7271	22.7
$T_{c,out}$	750	293.9	0.7984	60.1

Reference condition: $T_0 = 530\,°R$, $h = 127\,Btu/lb$, $s = 0.5963\,Btu/lb\,°R$

Figure 3-6. Air heat exchanger operating conditions.

The mean temperature difference is 200 °F.

$\Delta h_{hot} = 398.4 - 319.5 = 78.9\,Btu/lb$

$\Delta h_{cold} = 345.5 - 268.5 = 77.0\,Btu/lb$

$Airflow\ rate_{hot} = \dfrac{50{,}000\,Btu/hr}{78.9\,Btu/lb} = 633.7\,lb/hr$

$Airflow\ rate_{cold} = \dfrac{50{,}000\,Btu/hr}{77.0\,Btu/lb} = 649.4\,lb/hr$

The exergy available in the hot-air inlet is

$(125.1\,Btu/lb)(633.7\,lb/hr) = 79{,}300\,Btu/hr$

The exergy leaving in the hot airstream is

$$(74.9\,\text{Btu/lb})(633.7\,\text{lb/hr}) \;=\; 47{,}500\,\text{Btu/hr}$$

The exergy picked up by the cold airstream is

$$(90.8 - 46.3)(649.4) \;=\; 28{,}900\,\text{Btu/hr}$$

The exergy lost in transfer is

$$79{,}300 - 47{,}500 - 28{,}900 \;=\; 2900\,\text{Btu/hr}$$

which is 6% of the initial exergy in the hot-air inlet stream.

Condition B

$$T_{h,in} = 1150\,°\text{F} \qquad T_{h,out} = 850\,°\text{F}$$
$$T_{c,in} = 550\,°\text{F} \qquad T_{c,out} = 850\,°\text{F}$$

Performing the same calculations as those for condition A, the mean temperature difference between the airstreams is 300 °F, and the hot-airflow rate is 633.7 lb/hr, as before. The cold-airflow rate is

$$\frac{50{,}000\,\text{Btu/hr}}{76.0\,\text{Btu/lb}} = 657.9\,\text{lb/hr}$$

The exergy available in the hot-air inlet and exit are the same as those for condition A, but the exergy picked up by the cold airstream is

$$(74.9 - 33.8\,\text{Btu/lb})(657.9\,\text{lb/hr}) \;=\; 27{,}000\,\text{Btu/hr}$$

The exergy loss in the heat transfer is then

$$79{,}300 - 47{,}500 - 27{,}000 \;=\; 4800\,\text{Btu/hr}$$

or approximately 10% of the available exergy in the inlet hot airstream.

Condition C

$$T_{h,in} = 1150\,°\text{F} \qquad T_{h,out} = 850\,°\text{F}$$
$$T_{c,in} = 450\,°\text{F} \qquad T_{c,out} = 750\,°\text{F}$$

Performing the same calculations as those for condition A, the mean temperature difference between the airstreams is 400 °F, and the hot-airflow rate is 633.7 lb/hr, as before. The cold-airflow rate is

$$\frac{50{,}000\,\text{Btu/hr}}{75.0\,\text{Btu/lb}} = 666.7\,\text{lb/hr}$$

The exergy available in the hot-airstream inlet and outlet are the same as that for conditions A and B. The exergy picked up by the cold airstream is

$$(60.1 - 22.7\,\text{Btu/lb})(666.7\,\text{lb/hr}) = 25{,}000\,\text{Btu/hr}$$

The exergy lost in the heat transfer is

$$79{,}300 - 47{,}500 - 25{,}000 = 6800\,\text{Btu/lb}$$

which is 14% of the total available exergy in hot-air inlet stream.

The percent exergy loss as a function of the mean temperature for a constant heat-transfer load is plotted in Figure 3-7. These results indicate that the exergy loss through irreversible heat transfer may not be negligible and should be considered especially in energy-related systems that have a number of heat exchangers.

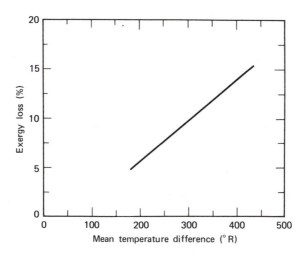

Figure 3-7.   Air heat exchanger exergy loss.

Example 3-2   Another important aspect of heat-exchanger performance is the method of determining the efficiency of a heat exchanger. Generally, heat exchangers are rated according to their heat-transfer effectiveness, which is (Kays and London 1958)

$$\epsilon = \frac{C_h(T_{h,in} - T_{h,out})}{C_{min}(T_{h,in} - T_{c,in})} = \frac{C_c(T_{c,out} - T_{c,in})}{C_{min}(T_{h,in} - T_{c,in})}$$

When the heat capacity (wc_p) of the two fluid streams are the same, the heat-transfer effectiveness can be replaced by the temperature effectiveness of the fluid stream

$$\epsilon_T = \frac{\text{Temperature change of fluid stream}}{\text{Maximum possible temperature change of stream}}$$

The importance of heat-exchanger performance for low-temperature cryogenic systems is highlighted by the research–development efforts that are applied to these components despite the fact that the temperature effectiveness of the fluid streams are around 96–98%. Figure 3-8 gives the conditions of a typical helium recuperative heat exchanger for a cryogenic refrigerator. The heat balance on this heat exchanger gives a heat-transfer effectiveness of

$$\epsilon = \frac{653.9 - 27.7}{688.3 - 52.5} = 0.985$$

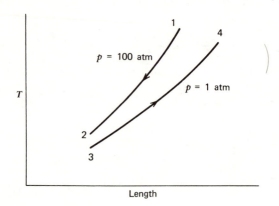

Length

Point	Pressure (atm)	Temperature (°R)	h (Btu/lb)	s (Btu/lb °R)	Ex (Btu/lb °R)
1	100	540	688.3	5.24	1250
2	100	36	52.5	1.73	2485
3	1	18	27.7	3.24	1635
4	1	522	653.9	7.46	0

Figure 3-8.   Helium recuperative heat exchanger.

The temperature-effectiveness values of the helium flow streams are

$$\epsilon = \frac{540 - 36}{540 - 18} = 0.966$$

$$\epsilon = \frac{522 - 18}{540 - 18} = 0.966$$

These high values of effectiveness indicate that the heat exchanger is designed well from a heat-balance standpoint. However, the picture is quite different when the second law and irreversibilities are considered. The exergy values shown in Figure 3-8 are calculated for a reference condition of helium at 340 °R and 1 atm pressure. The exergy balance using these values is

Exergy released $= Ex_4 - Ex_3 = 0 - 1635 = 1635\,\text{Btu/lb}$

Exergy absorbed $= Ex_2 - Ex_1 = 2485 - 1250 = 1235\,\text{Btu/lb}$

which shows that the exergy lost in the heat transfer is 400 Btu/lb. The exergy efficiency of this heat exchanger is then

$$\text{Exergy efficiency} = \frac{1635 - 400}{1635} = 0.755$$

These results show that this heat exchanger is not very efficient from an exergy standpoint. It is now desirable to ascertain the reason for the low exergy efficiency. To do this, the end temperature of the helium flow streams are varied to indicate the sensitivity of these areas. The exit temperatures of the warming and cooling streams are varied to provide temperature differences up to 30 °K. The exergy balance made for these conditions result in the curves shown in Figure 3-9. The change in exergy efficiency is much greater for a given change in temperature, ΔT, at the cold end than for the same temperature change at the warm end. Trepp (1961) has pointed out that efficient heat-exchanger design for low-temperature operation will have the temperature difference between the fluid streams decreasing with decrease in temperature to minimize irreversible heat-transfer losses. This can be done by increasing the mass flow

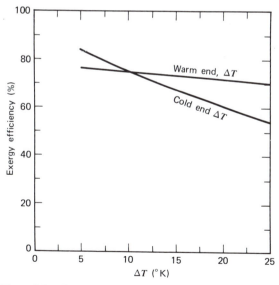

Figure 3-9. Sensitivity of heat exchanger end temperatures.

rate of the fluid stream in the low-temperature area through reducing the flow area. The pressure drop of the fluid stream will be higher because of this change. However, in many cases the increased power for the pump or blower to move the fluid will result in only a small increase in exergy loss compared to the gain by reducing the temperature difference in the fluid streams. Bejan (1978b) gives a thorough discussion of this relationship between irreversible heat transfer and pressure drop.

Another method of reducing the temperature difference in heat exchangers and the exergy loss associated with it is to increase the heat-transfer area. This can be done by building a larger heat exchanger or adding extended surface fins to the primary surface. These methods will result in increased first cost of the unit that must be considered in a trade-off for the most effective life-cycle cost and use of available energy sources.

The advent of heat pipes or thermosyphon tubes has provided a useful technique to improve the performance characteristics of heat exchangers. These devices do not eliminate the major source of exergy loss in air–air heat exchangers, the boundary layer, but they do remove some of the constraints on designing thermally efficient heat exchangers. For example, the heat exchanger shown in Figure 3-10 uses heat pipes or tilted thermosyphons to transfer heat from the hot gas stream to the cold airstream. This type of heat exchanger would be used as an air heater in a power plant or to remove waste heat from exhaust gases of a processing plant. Since the hot gas and the cold

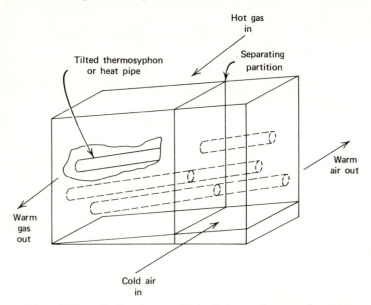

Figure 3-10.   Heat exchanger with heat pipes or thermosyphon tubes.

airstreams can have varying mass flow rates, pressure levels, allowable pressure drops and different temperature levels and radiation heat-transfer characteristics, the heat-transfer geometries should generally be quite different for overall optimum thermal performance of the heat exchanger. Conventional tubular-type heat exchangers impose a constraint in this respect because of the difficulty of internally finning the tubes, and they generally require cross-flow operation. The heat-pipe heat exchanger can be designed so that the fins installed at the evaporator end of the pipe are different than those installed on the condenser end of the pipe. Counterflow operation of the flow streams can be easily accomplished for most effective thermal performance.

However, the use of heat pipes in heat exchangers has little effect on the thermodynamic performance of the heat exchanger since the major thermal resistance to heat transfer is in the gas film boundary layer, which is not changed. Exergy losses in the heat-pipe heat exchanger will be reduced to the extent that the heat-transfer—pressure drop factor is better optimized because of the wider choice of heat-exchange surface configurations available by using heat pipes.

3-10 EXERGY LOSS IN COMBUSTION AND CHEMICAL PROCESSES

The greatest loss in available exergy is in the industrial sector of our economy. This is primarily because of the large amount of high-quality energy involved in combustion and chemical processing. The greatest gain in reducing exergy losses can be made in this area.

The exergy analysis of combustion and chemical processes is complicated by the fact that many individual processes are taking place at the same time and different species of matter are involved. The procedure is conducted by summing up the individual results. When the system undergoing chemical reaction is in equilibrium with the surroundings, the value of exergy is

$$Ex_i = (U + P_0 V_i - T_0 s_i) - (U_0 + P_0 V_0 - T_0 s_0) \qquad (3\text{-}10)$$

for conditions where the kinetic and potential energies can be neglected. Since $h_i = U_i + P_i V_i$, then

$$Ex_i = (h_i - h_0) - T_0(s_i - s_0)$$

which is the basic equation in the exergy system of analysis.

This latter equation is used for each species involved in the combustion or chemical process, and the total exergy is obtained by summing up the component values.

When this equation is used for combustion, the enthalpy used is the enthalpy of formation of the product of the reaction. In the reaction

$$2H_2 + O_2 = 2H_2O$$

The change in exergy is

$$Ex = (h_f^0)_{H_2O} - T_0(s_{H_2O} - s_{H_2} - s_{O_2})$$

The first term is the heat of formation of the combustion products. The second term in the preceding equation is the loss of exergy because of irreversibility in the process. This loss can be up to 50% of the available energy in the original fuel.

Since combustion usually involves the oxidation of several species in a fuel, the heat balance and exergy loss must be calculated for each species involved in the reaction. If we consider that there are m' species involved in the original fuel and m'' species involved in the products of combustion, the exergy in the original fuel will be

$$Ex = m'[(h_i - h_0) - T_0(s_i - s_0)]$$

and in the products of combustion

$$Ex = m''[(h_e - h_0)T_0(s_e - s_0)]$$

where the notation 0 denotes the reference condition for the species in question.

As an example of the use of the exergy equations in the combustion process, we consider the combustion of Ohio bituminous coal having the fuel analysis shown in Table 3-4. The calculations for the mass and energy balance in the combustion process are also shown on Table 3-4. These combustion calculations are typically performed for a specific fuel and operating condition during the design and performance analysis of each combustion system. Such calculations are now computerized.

The net result of these calculations shows that there is a 12.63% loss of heat input because of the unburned carbon and carbon monoxide and the heating of the moisture in the fuel and air during the combustion process.

The fuel heating value is 12,800 Btu/lb, which is the potential exergy from the fuel. Since 12.63% of this is lost through heat losses during combustion, the exergy available is reduced to

12,800(1 − 0.1263) = 11,183 Btu/lb

We next consider the irreversible losses in the combustion process itself. The exergy prior to the irreversible process is 11,183 Btu/lb. The exergy value of the products of combustion following the irreversible process is determined by evaluating the increase in entropy of each constituent and summing up these values. From Table 3-4 we have the following weight percent (neglecting the small amount of SO_2) of the flue gas:

CO_2	21%
O_2	4%
N_2	70%
H_2O	5%
	100%

Table 3-4   Combustion System Performance Calculations

Fuel: Bituminous coal, Ohio
Heat content: 12,800 Btu/lb
Fuel analysis: percent by weight as fired

Carbon	72.0
Hydrogen	4.4
Sulfur	1.6
Oxygen	3.6
Nitrogen	1.4
Water	8.0
Ash	9.0
Total	100.0%

Total air: 120%
Flue-gas exit temperature: 345 °F
Reference ambient temperature: 80 °F

Flue-gas losses:

CO_2	$9.75 \times 265 \times 0.058$	=	150 Btu/lb fuel
O_2	$7.2 \times 265 \times 0.0138$	=	26 Btu/lb fuel
N_2	$7.0 \times 265 \times 0.312$	=	579 Btu/lb fuel
H_2O in air	$8.1 \times 265 \times 0.0084$	=	18 Btu/lb fuel
H_2O in fuel, sensible heat			
	$8.1 \times 265 \times 0.0264$	=	57 Btu/lb fuel
H_2O in fuel, latent heat			
	$0.0264 \times 1040 \times 18$	=	494 Btu/lb fuel
Unburned carbon, 1% ash in Fuel:		=	155 Btu/lb fuel
Unburned carbon monoxide in flue gas,			
	$0.0011 \times 12 \times 10,200$:	=	135 Btu/lb fuel

Total flue-gas losses: 1614 Btu/lb fuel
Heat input of fuel: 12,800 Btu/lb
Total loss: 12.6%

Adapted from *Steam, Its Generation and Use* with permission from Babcock & Wilcox Company.

The entropy of the gas following the combustion process is evaluated by determining the entropy of each component at the adiabatic gas temperature and summing up these individual values. The temperature before the process is taken as room temperature, 530 °R, and the adiabatic gas temperature is

$$T = \frac{(12,800 \text{ Btu/lb fuel heat value})(1/12.66\text{-fuel:gas ratio})}{(0.27 \text{ Btu/lb °R average specific heat})}$$

$$T = 3744 °R \text{ or } 3284 °F$$

where the fuel:gas weight ratio is obtained from Table 3-4 for the wet flue gas and the specific heat is averaged over the temperature range.

The entropy of the fuel in its initial condition is calculated on the basis of one pound of fuel:

Constituent	Weight (lb)	Entropy
Fuel		
C	0.72	0.11 = 0.082
H_2	0.044	16.1 = 0.687
O_2	0.036	1.53 = 0.055
N_2	0.014	1.63 = 0.023
H_2O	0.08	2.51 = 0.200
Air		
O_2	2.45	1.53 = 3.75
N_2	9.21	1.63 = 15.060
		19.857 Btu/lb fuel °R

The entropy of the flue gas after combustion at 3744 °R is

Constituent	Weight (lb)	Entropy
CO_2	0.21	1.69 = 0.356
O_2	0.04	2.02 = 0.081
N_2	0.70	2.16 = 1.513
H_2O	0.05	3.54 = 0.177
		2.127 Btu/lb flue gas

To obtain the latter value on the same pound-fuel basis,

$$2.127 \text{ Btu/lb gas} \frac{12.66 \text{ lb gas}}{1 \text{ lb fuel}} = 26.928 \text{ Btu/lb fuel}$$

The exergy loss through irreversibilities in the combustion process is then

$$Ex_{loss} = T_0(S_{gas} - S_{fuel}) = 540(26.928 - 19.857)$$
$$= 3818 \text{ Btu/lb}$$

The exergy available after the combustion process is then

$$11{,}183 - 3818 = 7365 \text{ Btu/lb}$$

The total exergy loss during combustion is

$$12{,}800 - 7365 = 5435 \text{ Btu/lb}$$

or the exergy available after the combustion process is only 57.5% of the exergy available in the initial fuel.

One aspect of this problem is what can be done to reduce this souce of exergy loss. One interesting example concerns studies that are in progress to reduce the formation of nitrogen oxides in the combustion process to minimize atmospheric pollution. An approach described by McKay (1977) would result in the maximum combustion temperature being reduced while maintaining a good heat-transfer rate to the working fluid. This process has the additional benefit of reducing the exergy loss incurred in the combustion process. The reduction in exergy loss achieved by reducing the combustion temperature can be shown by the following simple example.

Assume that the conventional combustion temperature is $3960\,^\circ R$, which is reduced to $2460\,^\circ R$ by using the concentric-tube burner described by McKay. For simplicity, we assume that the burners are used to generate saturated steam at 600 psia ($946\,^\circ R$ saturated temperature). The ambient temperature of $530\,^\circ R$ is used as the exergy reference condition.

The exergy required to perform the boiling per unit of heat required is

$$Ex_{required} = 1 - \frac{T_0}{T_s} = 1 - \frac{530}{946} = 0.44$$

The exergy available from the conventional burner is

$$Ex_{available} = 1 - \frac{530}{3960} = 0.87$$

In this case 49% of the exergy or available work from the fuel is lost in the heat-transfer process. The exergy available from the concentric-tube burner is

$$Ex_{available} = 1 - \frac{530}{2460} = 0.78$$

In this case only 44% of the exergy of the fuel is lost, resulting in a 10% reduction in the exergy loss during the combustion process. Since this saving is accomplished with high-quality heat, the benefit to most energy related systems using combustion of fossil fuels can be significant.

From a thermodynamic point of view the combustion process offers the potential for significant savings in exergy loss and the opportunity to improve the efficiency of energy-intensive systems. The combustion process is complex, and in most cases the designs are based on empirically developed data. However, recent progress in furnace and combustion chamber modeling (Steward and Guruz 1974) is providing the tools to understand and improve combustion systems.

Since the heat of combustion in a gas stream is usually used to heat a working fluid such as steam, the potential improvements in exergy efficiency during combustion must consider any effect on the heat transfer in the subsequent processes.

3-11   EXERGY LOSS IN FLUID FLOW

Loss in exergy occurs in fluid flow systems as a result of friction, viscosity, heat-leak, momentum, and gravity effects. The degree to which these irreversible losses are considered in a system depends on the relative effect of these losses on the overall system efficiency. In many energy systems the exergy losses through combustion, heat transfer, and machine inefficiencies (turbines or compressors) are dominant over the fluid-flow losses. In these systems fluid-flow exergy losses can sometimes be neglected. However, there are many industrial systems that involve little thermal energy where fluid-flow losses dominate. Typical systems of this type include hydroelectric power plants, water pumping and distributing systems, coal slurry and oil pipelines, and air-distribution systems in mines.

To generalize the loss of exergy in a fluid-flow system, we consider the exergy of the fluid at point i as

$$Ex_i = U_i - U_0 - v_i(P_i - P_0) + mg(z_i - z_0)$$
$$+ \tfrac{1}{2}m(V_i^2 - V_0^2) - T_0(s_i - s_0) \tag{3-11}$$

where

U = internal energy of fluid
v = volume per unit mass
P = pressure
m = mass flow rate
g = acceleration of gravity
z = elevation
V = velocity
s = entropy
0 = reference exergy condition

In the general case the reference exergy condition will be the condition of the fluid at rest at sea-level ambient conditions. However, for many exergy analyses it may be more convenient to use some other reference, such as the outlet duct of a hydroelectric power plant when performing an exergy analysis of the plant itself.

In proceeding from point i to point j under real conditions the fluid will change in pressure, velocity, and possibly elevation. The entropy of the fluid will also change as a result of heat generated by viscous dissipation in friction. These losses are calculated in the conventional first-law manner using a thermal–mechanical energy balance. When the conditions at point j are determined, the value of exergy at point j can be evaluated using Eq. (3-11). The total exergy loss in the fluid between points i and j can be obtained by taking the difference of the two values. In many cases where the exergy losses are high it is desirable to separate the individual losses so that improvements in the system efficiency can be directed in the most effective manner. In this case the specific individual exergy losses in the process are calculated using the following:

Internal energy	$Ex_{loss} = m(U_j - U_i)$
Friction	$Ex_{loss} = m(P_j v_j - P_i v_i)$
Elevation (potential energy)	$Ex_{loss} = m(g/g_c)(z_j - z_i)$
Momentum (kinetic energy)	$Ex_{loss} = \frac{1}{2}m(V_j^2 - V_i^2)$

As an example, consider a water pump supplying 2000 lb/hr of water to a tank 50 ft above the supply level. The pump provides the water at 200 psia and a velocity of 50 ft/sec while the water leaves the pipe in the tank at 10 ft/sec and atmospheric, sea-level pressure. This example has a velocity change (kinetic energy), a pressure change (friction), and an elevation change (potential energy). The total exergy loss in the system and the work required by the pump to overcome these losses are calculated directly as work-loss terms.

$$\text{Friction } Ex_{loss} = \frac{2000 \text{ lb/hr } (50 - 14.7 \text{ psia}) \, 144 \text{ psia/psfa}}{62.4 \text{ lb/ft}^3}$$

$$= 162{,}920 \text{ ft-lb/hr}$$

$$\text{Potential-energy } Ex_{loss} = 2000 \text{ lb/hr} \left(\frac{32.2 \text{ ft/sec}^2}{32.2 \text{ ft/sec}^2} \right) (50 \text{ ft-0})$$

$$= 100{,}000 \text{ ft-lb/hr}$$

$$\text{Kinetic-energy } Ex_{loss} = \frac{2000 \text{ lb/hr } [(50 \text{ ft/sec})^2 - (10 \text{ ft/sec})^2]}{2(32.2 \text{ ft/sec}^2)}$$

$$= 74{,}530 \text{ ft-lb/hr}$$

The total exergy loss is 337,450 ft-lb/hr or 0.17 hp for a 100% efficient motor and pump.

3-12 EXERGY AS A STANDARD FOR ENERGY-SYSTEMS COMPARISON

To assess the advantages of one energy system or type of system over another comparisons are made on the basis of system efficiency and cost. The use of the conventional energy-balance efficiency, where heat energy and electrical energy are considered equal, is unrealistic and misleading. The use of the exergy efficiency, which is based on the second law of thermodynamics, is a realistic method of making comparisons between systems of the same type and between different types of system. In this section the application of the exergy method to energy-systems comparison is reviewed.

First let us look at an extreme example of the inability of the heat-balance efficiency to give realistic results. In the Carnot cycle heat must be supplied to the system at one temperature, and heat must be rejected to the surrounding heat sink at a lower temperature for the cycle to operate. The system is penalized for the heat that is rejected to the surrounding heat sink since that rejected heat does not perform work. On the other hand, the throttling process is an isenthalpic process in which no heat is lost; therefore, in a heat-balance analysis this process is an efficient one. Actually, from a work-efficiency stand-point the Carnot is the best thermodynamic process whereas the throttling process, because of the loss of work in expanding the high-pressure gas, is one of the least efficient processes that we have in energy systems. The efficiencies of these two processes are put into the correct perspective when the exergy comparison using the second law of thermodynamics is applied.

The basic exergy equation [Eq. (1-2)] considers the terms for the most common forms of work used in energy systems. These include gravity head, flow momentum, electrical potential, high pressure, high fluid temperature, and high surface temperature. There are other less common forms of work that can also be integrated into the exergy equation by considering the available work under a given condition relative to the condition at rest in the ambient surrounding.

Since work of all forms can be put into a single reference, direct comparison between types of energy system can be realistically performed. The form of work used as a reference is not important as long as the comparison covers the same boundaries of the system. For example, when comparing an oil-fired gas-turbine-driven compressor system with an electric motor-driven compressor, the irreversible losses in the generation of electricity for the motor from the fossil fuel or nuclear power must be considered to make a valid comparison.

CHAPTER **4**

Analytical Procedure

In this chapter the basic method for calculating exergy losses is discussed and the block method of exergy analysis presented.

The basic calculation of exergy losses is shown with the aid of $T-s$ diagrams that clearly indicate the exergy losses qualitatively and quantitatively for each process described. These basic calculation procedures are used later in block form to provide the basic for a block method of exergy calculations.

In the block method of exergy analysis the equations for calculating the exergy change and exergy loss in each process are presented in a block form. The input and output of each block involves the pertinent performance parameters of the energy system, including the values of the exergy at the beginning and the end of the process under consideration in the block. An example is given for each process block.

The procedure for using the block method of exergy analysis is described in this chapter. Examples of the application of the block method are given in Chapter 5.

Additional blocks for combined processes, detailed processes, or processes not covered in this chapter can be developed using the information provided in Chapter 3.

4-1 BASIC CALCULATION OF EXERGY LOSS

The increase in entropy production of a process in the presence of irreversible effects can be used to calculate the loss of useful work (exergy loss).

If a working fluid receives heat in a reversible process ($1-2$ in Figure 4-1), the maximum useful work from this heat is accomplished in a reversible cycle if the cold sink is the surrounding medium with essentially a constant temperature

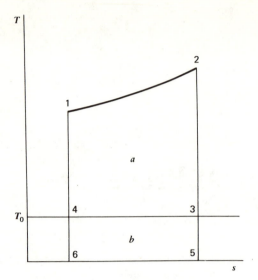

Figure 4-1.   Ideal heat transfer.

T_0 and the processes of expansion and compression proceed along two equidistant isentropes in which all adiabatic conditions are satisfied. This maximum quantity of possible useful work in a cycle for a given heat source is called the *ideal* or *maximum available exergy* (or *work*) of the fluid or body. In the *T–s* diagram it is obvious that the exergy is measured by the area bounded by the curve of the heat-supply process, with the ordinates and the isotherm having a temperature of the surrounding medium. The maximum exergy is greater when the heat is supplied along the curve $(1-2)$ at a greater temperature.

For the ideal case (Figure 4-1), the exergy, according to Eq. (3.1), is

$$Ex = q_0 = q_1 - q_2 = (a + b) - b = q_1 - T_0(s_3 - s_4)$$

where:

q_0 = maximum possible useful work

q_1 = supplied heat

q_2 = rejected heat

$s_3 - s_4$ = decrease in entropy of cold sink

$s_2 - s_1$ = equal to $s_3 - s_4$ and absolute value of increase in entropy of warm source

The change of entropy in such a system composed of reversible processes is zero. This ideal case does not exist in nature, and the most common deviations from the pure reversible process or cycle are described in the following paragraphs.

We calculate the exergy loss during heat transfer between a warm source of heat at T_x and a working fluid (Figure 4-2). We let the same quantity of heat be transferred first in a reversible manner and then in an irreversible manner.

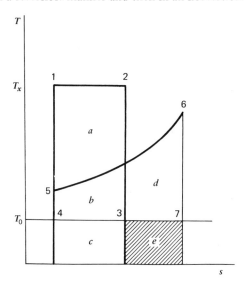

Figure 4-2. Heat transfer, infinite heat source to fluid.

The warm source is assumed to have a large heat capacitance so that removing heat from it does not change its temperature. For the reversible case, the working fluid receives heat along line 1–2 and the exergy (or available work) is 1–2–3–4–1. For the irreversible case, the corresponding line is 5–6 and the exergy is 5–6–7–4–5. In the first case the transfer of heat is assumed to take place with no temperature difference between the fluids. Although this is an ideal case and not possible in a real heat-exchange system, it is essentially what is assumed when heat-exchange performance is calculated by the heat-balance or first-law method. In the irreversible case a temperature difference exists between source 1–2 and working fluid 5–6, and it is impossible to reverse the process because of the finite-temperature difference. It should be noted that if the heat were supplied along path 5–6, cycle 5–6–7–4–5 would be reversible. Hence this latter case is reversible internally but irreversible externally.

The exergy loss from the heat transfer between the warm source and the working fluid is calculated in the following manner. The exergy available during reversible heat transfer is

$$Ex_{rev} = q_1 - q_2 = (a + b + c) - c = a + b$$

The exergy available after irreversible heat transfer is calculated for the case of heat supplied along lines 5–6:

$$Ex_{\text{irreversible}} = q_1 - q_2 = (b + d + c + e) - (c + e) = b + d$$

The loss of exergy is then

$$Ex = Ex_{\text{reversible}} - Ex_{\text{irreversible}} = (a + b) - (b + d) = a - d$$

Since q_1 in both cases is the same, $a + b + c = b + e + d + c$ and $a = d + e$. Also

$$Ex = a - d = d + e - d = \text{Area } e = T_0(3-7)$$

The change in entropy of the system during the presence of heat transfer is now calculated. For the reversible adiabatic lines (6–7 and 4–5), there is no change in the entropy of the working fluid. The absolute entropy change of the warm source is

$$\Delta s_{\text{source}} = \left| \frac{a + b + c}{T_x} \right| = \text{section } 1-2$$

where T_x is the temperature of the warm source. The increase in entropy of the working fluid is

$$\Delta s = \text{section } 4-7$$

Therefore, the increase in the entropy of the system is

$$\Delta s_{\text{system}} = s_{\text{fluid}} - s_{\text{source}} = \text{section } 4-7 - \text{section } 1-2$$

$$= \text{section } 3-7$$

This results in an exergy loss of

$$Ex_{\text{loss}} = T_0(\text{section } 3-7) = T_0 \Delta s_{\text{system}}$$

which is the same equation calculated by the work-area method previously.

Therefore, the loss in exergy from irreversible heat exchange is determined as the product of the temperature of the surrounding medium and the increase in entropy of the system because of the irreversible transfer of heat in it.

The exergy loss in the case of irreversible heat exchange between the working fluid and a cold sink is represented by Figure 4-3. Let the heat transfer from the warm source be reversible in this case and consist of an infinite series of heat sources along line 1–2. For calculating the loss of exergy, we consider the initial case when heat exchange occurs by a reversible path along line 5–6 and then by an irreversible path along line 3–4 through a thermal resistance. In this latter case the heat exchange is considered as internally reversible but externally irreversible. It is necessary to note that in this case the heat exchange differs from the previous case in that the relationship of the irreversible heat exchange to the reversible is derived for a different quantity of heat.

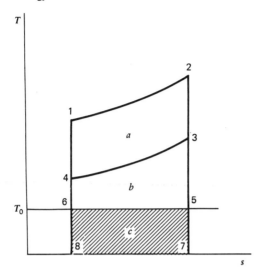

Figure 4-3. Heat exchange between same fluid.

So for this case,

$$Ex_1 = q_1 - q_2 = (a + b + c) - c = a + b$$

$$Ex_2 = q_1 - q_2 = (a + b + c) - (b + c) = a$$

$$Ex_{loss} = Ex_1 - Ex_2 = (a + b) - a = b$$

The change in entropy of the system during irreversible heat exchange is determined by process 3–4. The absolute value of the decrease in entropy of the working fluid is

$$\Delta s_{fluid} = (s_3 - s_4)$$

The increase in entropy of the cold sink for a constant temperature T_0 and a given quantity of heat as measured by the area $b + c$ is

$$\Delta s_x = \frac{b + c}{T_0}$$

The change of entropy in the system is

$$\Delta s_{system} = \frac{b + c}{T_0} - (s_3 - s_4)$$

The loss of exergy as measured by the area b is represented by

$$Ex_{loss} = (b + c) - c = (b + c) - T_0(s_3 - s_4)$$

or

$$Ex_{loss} = T_0 \left[\frac{b+c}{T_0} - (s_3 - s_4) \right]$$

But the term in the bracket represents the net exchange of entropy in the system. Therefore,

$$Ex_{loss} = T_0 \, \Delta s_{system}$$

Figure 4-4 shows a more general case of irreversible heat transfer. Let such a heat transfer occur from fluid I along curve 1−2 to fluid II along curve 3−4. For simplicity, we consider that the entropy s_2 of fluid I at the end of the process is equal to entropy s_3 at the beginning of the process in fluid II. The heat capacities of the two fluids are assumed to be the same. When two fluids have the same heat capacitance, they can be plotted on the same $T-s$ diagram. If the two fluids have different heat capacitances, they can only be plotted together on the $T-s$ diagram if an equivalence factor is applied to one of the fluids.

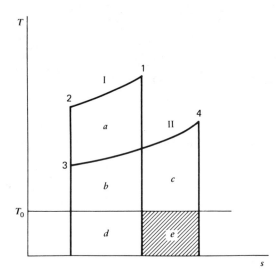

Figure 4-4.   Heat exchange between two different fluids.

For the change in the conditions of the two fluids to be represented in the $T-s$ diagram as reversible and the quantity of heat transferred as shown by the areas under the process curves, it is necessary to consider that the heat transferred from fluid I to fluid II flows through a thermal resistance so that there is thermal uniformity at the boundaries of the fluids. Under this condition the quantity of heat leaving fluid I is equal to the quantity of heat entering fluid II. The areas representing the quantity of heat exchanged are

$$q = a + b + d = b + c + d + e$$

or after simplifying:

$$a = c + e$$

The available exergy in fluid I is

$$Ex_I = a + b$$

and in fluid II:

$$Ex_{II} = b + c$$

The loss of exergy as a result of the irreversible heat transfer is then

$$Ex_{loss} = Ex_I - Ex_{II} = (a + b) - (b + c) = a - c$$

Since $a = c + e$, we get

$$Ex_{loss} = (c + e) - c = e$$

Therefore, the loss of exergy is measured by the area e, the value of which is represented by the product

$$T_0(s_4 - s_1)$$

but $s_4 - s_1$ is not different from the change in entropy of the system that is produced by irreversible heat transfer. Therefore,

$$Ex_{loss} = T_0(s_4 - s_1) = T_0 \Delta s_{system}$$

that is, we obtain the same results here as those obtained for the previous cases.

Another reason for exergy loss is the friction in a working fluid that arises both at the boundary of the fluid in contact with surfaces and within the fluid stream as a result of turbulent motion of the fluid. The friction is an irreversible process since the work of friction changes into heat, and even in the case where the heat is completely absorbed by the working fluid, the mechanical energy expended in the friction cannot be fully recovered, according to the second law of thermodynamics.

We consider the exergy loss from friction in a working fluid during adiabatic expansion in a turbine as an example. In a turbine the moving gas proceeds from a high velocity, and the friction encountered in the process is the reason for the loss of mechanical energy. The process we examine is shown as $1-2-3'-4-1$ in Figure 4-5. Point 3 is primed to indicate that it is an ideal point (based on reversible assumptions). For considering the loss in exergy because of friction we calculate how the exergy changes in the cycle if the adiabatic expansion process

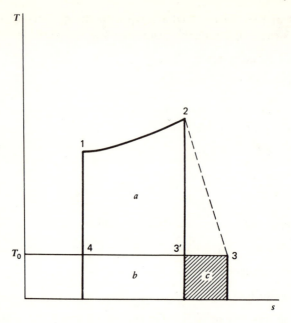

Figure 4-5.   Nonisentropic expansion, typical internal system exergy loss.

occurs irreversibly (with friction). Such a process occurs with an increase in entropy, and the final state is shown by point 3 lying on the same isotherm but with a greater value of entropy than at point 3′. It i͟ necessary to consider the process change of conditions between point 2 and point 3. This process is irreversible and theoretically cannot be represented as a line on the T-s diagram. Since only the end steady-state conditions are real, we show this process by a dotted line. The area under this line has no real significance to the process. The loss of exergy can be presented as follows. For the case with reversible expansion (2–3′), the exergy is

$$Ex_{\text{reversible}} = (a+b)-b = a$$

For the case of the irreversible process (2–3), the exergy of the system is determined as the maximum possible useful work that is equal to the difference in the quantity of heat received from the warm source and that given up to the cold sink:

$$Ex_{\text{irreversible}} = q_1 - q_2 = (a+b)-(b+c) = a-c$$

Then the loss of exergy is

$$Ex_{\text{loss}} = Ex_{\text{reversible}} - Ex_{\text{irreversible}} = a-(a-c) = c$$

The area c is represented by the product of the temperature of the surrounding medium T_0 and the line $s_3 - s_3'$, which measures the change in entropy of the system.

Actually, the entropy of the warm source from which heat is transferred to curve 1–2 decreases by $s_2 - s_1$ (absolute value), and the entropy of the cold sink to which heat is transferred along the line 3–4 increases by $s_3 - s_4$. Since the heat transferred in the two cases is reversible, the change in the entropy of the cold-sink heating is equal to the change in the entropy of the working fluid in the process of absorbing and giving up heat.

The entropy of the working fluid within the process loops does not change. Therefore, the overall increase in entropy of the system is

$$\Delta s_{system} = (s_3 - s_4) - (s_2 - s_1) = s_3 - s_2 = s_3 - s_3'$$

and the loss of exergy in the cycle because of friction is

$$Ex_{loss} = \text{Area } c = T_0 \, \Delta s_{system}$$

Thus in all cases it is found that the loss of exergy caused by irreversibility is calculated as the product of the absolute temperature of the surrounding medium and the change in entropy of the system. The calculations can be reduced to adding up all the irreversibility-induced entropy increases and multiplying by the surrounding temperature, but this would defeat one of the purposes of the exergy method of analysis, which is to show the extent and location of the individual exergy losses.

The processes that were shown previously are typical for industrial energy plants and electrical power plants. There are many other processes involving irreversibility where exergy losses occur, but they are less common in our industrial system and involve a relatively negligible loss of exergy. An exception is the combustion or chemical conversion process, which is the subject of a separate section in this book.

Exergy losses also occur in open systems where a mass containing energy in the form of heat, pressure, elevation, and so on is mixed with the surrounding medium. In such a process all the available work or exergy in the mass is lost.

4-2 BLOCK METHOD OF EXERGY ANALYSIS

One method of performing the exergy analysis for a complex system is to use a block method comprised of a flowchart of the system and individual calculating procedures for various processes. This block method provides a technique for forming the analytical computations to give any degree of refinement desired in the results. The method also provides a simple bookkeeping procedure for the results. Once the block method is set up in a flowchart, the computations can be readily made by engineering aides without the necessity to fully understand the exergy method of analysis or the second law of thermodynamics. The block method and the flowchart also form the basis for programming of computer computations, especially for more complex systems. The block calculations can be set up as subroutines to be called out as required for analysis of a system.

The exergy-computation procedure for many common processes have been developed in block form, and these process blocks are listed in Table 4-1. These process block procedures are given in Appendix A.

Each block shows the $T-s$ diagram and the equations for calculating the exergy loss in the process. Sample calculations of the exergy values and the exergy losses are given for each process block in Appendix A. These blocks cover many of the technical processes that will contribute to irreversible losses in energy systems. The exergy change and the loss of exergy in other processes not covered by these blocks can be developed from information in reports or thermodynamic books, some of which are listed in the bibliography. For example, a block can be developed from performance equations reported by Gidaspow (1978). Although

Table 4-1   Exergy-Analysis Blocks

Block number	Process	Component
1	Combustion, overall	Furnace, heater
2	Combustion, chemical energy to heat energy	Furnace, heater
3	Combustion, heat transfer gas to working fluid	Boiler, heater
4	Frictional loss, vapor or gas flow	Ducts or pipes
5	Frictional loss, liquid flow	Pipes
6	Heat loss, vapor or gas flow	Ducts or pipes
7	Heat loss, liquid flow	Pipes
8	Flow heat and frictional loss, vapor or gas flow	Ducts or pipes
9	Flow heat and frictional loss, liquid flow	Pipes
10	Pumping, liquid	Liquid pump
11	Expansion with work, vapor	Turbine, engine
12	Expansion with work, gas	Turbine, engine
13	Condensation, saturated vapor to liquid	Condenser
14	Heat transfer, different fluids	Heat exchanger
15	Heat transfer, same fluids	Heat exchanger
16	Mixing, different temperatures	Piping system
17	Compression, gas or vapor	Compressor

these results require validation and further refinement by additional system tests, the use of such a block can show the benefit of further development but would be limited to exploratory or preliminary system studies until validated performance is available.

In other words, the real effect of a process variation or improvement on the overall system efficiency can be determined, although technically the improvement is not proven. An evaluation of this type can provide the basis for management decisions as to whether development expenditures are warranted.

The procedure for using the block method of analysis starts with a system schematic and a flowchart of the computations required to analyze the performance of the system. The flowchart can include conventional component calculations and exergy calculations as required to meet accuracy requirements and the time and manpower constraints.

The first approximation in a systems analysis is to evaluate only the major components, and these are considered to have ideal performance (100% efficiency); that is, all processes are reversible and adiabatic. These analyses are useful only for general evaluation of the system, and in most cases these ideal results will be erroneous and misleading if they are used for system or performance comparisons, especially if efficiencies are compared.

In most cases this ideal reversible analysis of a cycle or system is not done, but the initial analysis is performed for a theoretical cycle that is neither ideal nor actual. In general, this theoretical cycle considers that major components, such as compressors, turbines, and pumps, have a certain empirical isentropic efficiency developed from tests of similar units. Although this is the common basis to which the exergy efficiency is compared, there is some question of the validity of such comparisons (Appendix B-4). On the other hand, this comparison is made under similar system operating conditions, and the direct benefit of the exergy analysis can be shown, although the reference base may not be the maximum capable performance criterion. The use of the ideal reversible, adiabatic cycle cannot be used as a base reference for the exergy analysis since the conditions before and after the process are generally not the same.

Typical systems analyses considering irreversible losses in the block method are shown for three different degrees of computational refinement. The first flowchart (Figure 4-6) is a preliminary analysis of the system shown in Figure 4-7, in which only the major components in the system are considered, but in this case the irreversible losses resulting from inefficient component operation are considered. The second flowchart (Figure 4-8) is used with the schematic in Figure 4-9 and involves the analysis of additional losses associated with the secondary components in the system. This provides a more realistic evaluation of the system efficiency but does involve additional computations in both the heat balance and the exergy analysis. The third flowchart (Figure 4-10) includes all potential sources of exergy loss in the system shown in Figure 4-11

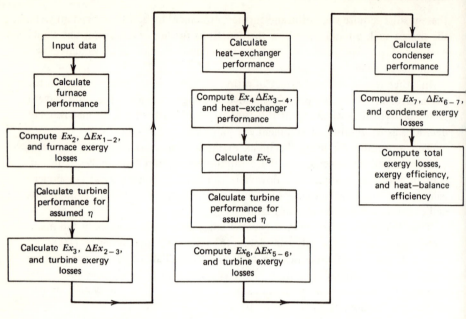

Figure 4-6.   Preliminary exergy-analysis flowchart.

and provides an accurate assessment of the system and individual component efficiencies. This latter analysis can be used to make realistic design improvements in the system since the relative magnitude of the exergy losses in the different locations of the system are accurately known.

The flowchart in Figure 4-10 is simplified to a single case. Generally, iterative loops are included in the program to vary system design parameters so that optimum design and operating conditions can be developed for maximum system exergy efficiency. Direct comparison of the performance efficiencies of different systems can also be made using the refined approach to provide realistic results that are more accurate than comparisons using the conventional heat-balance approach. This latter, refined approach for system analysis will require more extensive calculations; however, with the availability of computer programs, the benefits far outweight the added analytical and computational effort.

As shown by the flowcharts, the computation blocks for the exergy analysis are inserted into the flow of the conventional heat-balance computations at the desired points of exergy- and exergy-loss evaluation. The output of the exergy computation block are values of the outlet exergy, the change in exergy in the process, and the exergy loss in the process. The outlet exergy and the change in exergy values are used in subsequent calculations to make an exergy balance

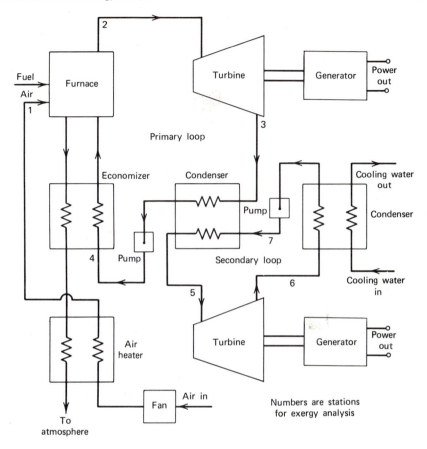

Figure 4-7. Preliminary exergy-analysis diagram.

and to assure that all the exergy in the initial energy source is accounted for by either loss or work output. The exergy losses from each process in the system are tabulated to show the location and the value of each exergy loss. This table is used to evaluate potential system improvements. The exergy losses are also totaled and the overall system efficiency determined from

$$Ex = \frac{Ex_{available} - Ex_{loss}}{Ex_{available}}$$

This efficiency can be used for comparison of different energy systems to show which type and configuration of system is more efficient in its use of the work available in the energy source.

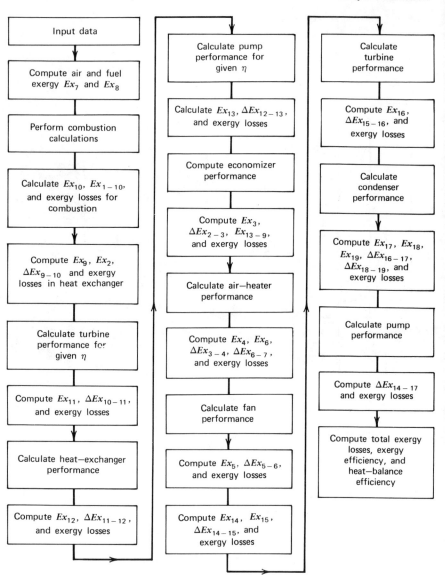

Figure 4-8.   Intermediate exergy-analysis flowchart.

4-3   PRESENTATION OF RESULTS — TABLES AND CHARTS

The basic information from an exergy systems analysis is tabulated numerical data as shown in Table 4-2, where the individual exergy losses are categorized

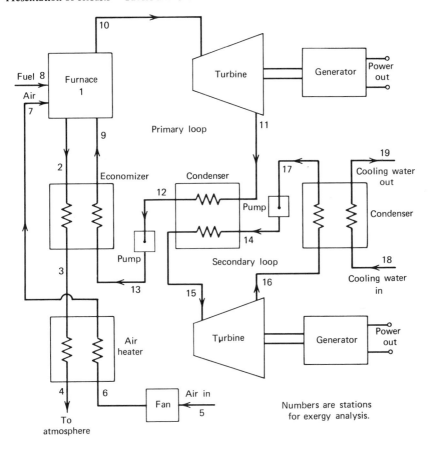

Figure 4-9.   Intermediate exergy-analysis diagram.

relative to the cause of the exergy loss. This breakdown is important in exergy-conservation studies since the cause of the exergy as well as its magnitude in the system will determine how effective system design or operation changes can be in energy conservation. The tabulation of exergy data in the form shown by Table 4-2 has the flexibility required for large and complex energy systems as well as simple small systems. This tabulation can also be easily generated by computer, and the data can be readily grouped for computer analysis of the results.

Although Table 4-2 shows the detailed performance characteristics of a system based on an exergy analysis, further manipulation of the data is usually required to present the results to other engineers, to company management, and to the general public. For example, the exergy losses in a system that are

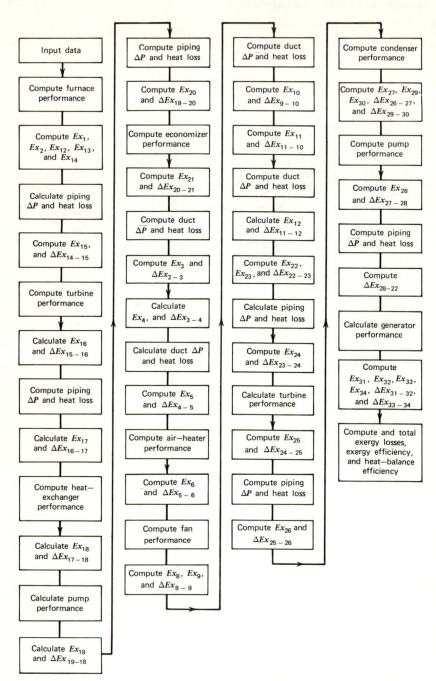

Figure 4-10. Detailed exergy-analysis flowchart.

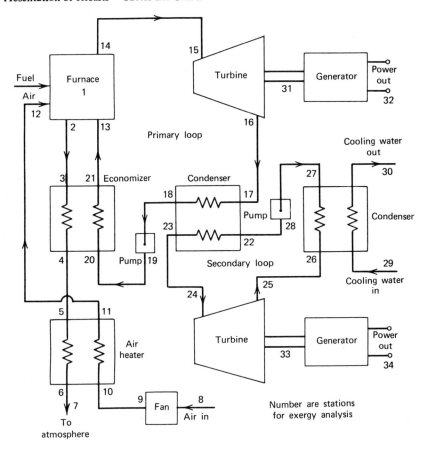

Figure 4-11.   Detailed exergy-analysis diagram.

attributed to heat transfer can be identified for special evaluation by heat-transfer engineers to determine ways in which these exergy losses can be minimized. A tabulation of the exergy losses resulting from rejection of mass or heat to the surrounding environment would be of special interest to researchers involved in energy recovery. The magnitude and locations of these exergy losses and the temperature levels at which they occur can be tabulated for use by engineers evaluating energy-recovery possibilities. It should be noted that the recovery of low-quality energy may not be economically practical unless a use for the recovered energy is found that matches the available energy well.

The overall effectiveness of the exergy analysis is usually shown by comparison with the conventional energy-balance method. Tabulated values of efficiency are common, but bar charts are also frequently used. The decision

98

Table 4-2 Exergy-Analysis Summary Form

Station or Component	Exergy	Change in Exergy	Heat Transfer	Useful Work	Chemical Change	Heat Transfer	Friction	Machine Inefficiency	Heat Rejection	Total	Total Exergy Loss (%)

of whether to use tables or charts will be dictated by the information to be presented and the audience to which it is being presented.

Diagrams are used to present the performance characteristics of systems so that comparisons and conclusions can be based on visual effects rather than numerical data. Diagrams and charts of energy-system characteristics can be used to perform two different functions. One case involves the presentation of detailed system performance to demonstrate the design and operating characteristics of the system to technically trained personnel. These diagrams and charts are used generally to supplement tables of numerical data and to simplify a technical presentation or report. The other case involves the presentation of system performance characteristics in such a manner that the advantages, disadvantages, or relative values of certain parameters are obvious from the diagram alone and nontechnical personnel can assess the impact of the performance characteristics being shown. Both types of diagram are useful in exergy-systems analysis, and the application of several proposed diagrams are discussed in this section.

Several diagrams have been proposed to visually show exergy-analysis results. These visual aids are useful for explaining the exergy-analysis results and how they compare with the other types of analysis. The temperature–entropy diagram (Figure 1-21) is widely used to show the effect of irreversibilities in processes and systems. As described in Section 4-1, the $T-s$ diagram will show the reversible and irreversible work for processes and cycles by the areas bounded by temperature and entropy lines. Unfortunately, the $T-s$ diagram and entropy itself have been difficult subjects to present even to engineers, and the majority of engineers proceed to forget the second law because of this problem. The $T-s$ diagram is useful for presenting the technical performance of a process or cycle to engineers, but more suitable diagrams described later are available for presentations of the exergy analysis to the general public. The $T-s$ diagram is also limited to a single working fluid and is not directly applicable to combined-cycle systems.

An enthalpy–entropy diagram is very useful in showing system performance characteristics and can be used to show the relationship of exergy to enthalpy and entropy differences in a system.

The exergy–enthalpy diagram was proposed by Brodyanskii (see Appendix B-2) to show the magnitude of exergy losses by areas enclosed by the exergy and enthalpy lines for a system. Since exergy is defined as

$$Ex = h - T_0 \, \Delta s$$

the areas enclosed by the exergy and enthalpy lines represent exergy loss because of irreversible work. This type of diagram is useful for comparing different configurations of the same system since the relative magnitudes of the exergy losses and available work are shown together on the same diagram. The exergy–

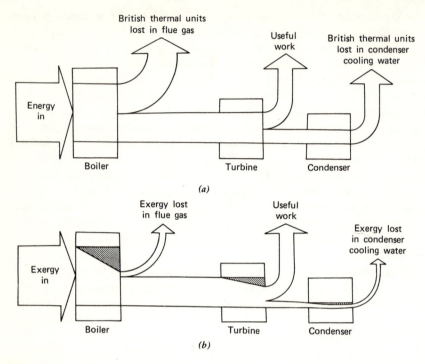

Figure 4-12.   Energy (*a*) and exergy (*b*) flow diagrams. (cross-hatched areas are internal exergy losses).

enthalpy diagram is useful for presenting technical data to engineers but would have the same limitation that the $T-s$ diagram has in regard to the nontechnical general public.

A useful diagram for presenting exergy-analysis results to the nontechnical general public is the one proposed by Baehr (1962) and Brodyanskii (see Appendix B-9) and represented by Figure 4-12. This diagram is a flowchart of the available work as it progresses through system components. The magnitude of the useful work and exergy losses in the components are visually shown by the relative widths of the work flow lines. This diagram shows the magnitude and locations of the exergy losses in a clear and effective manner. The presentation of exergy-analysis results is even more effective when this exergy flow diagram is used in conjunction with a similar energy flow diagram developed from a first-law heat balance of the system. A comparison of these two flowcharts (Figure 4-12) shows how significantly the exergy method influences the location and magnitude of the inefficiencies in energy systems. Since the goal of energy-conservation measures is to increase the useful work from that available in the initial energy resource, the exergy diagram shows the true distribution of losses and avoids the erroneous information given by the energy-balance method. The flowchart diagram can be refined in several ways to more fully present the results of the exergy analysis. For example, the quantitative values of the individual losses making up the total component loss can be listed in the chart shown in Figure 4-12 along with the type of loss. These data will be useful in discussion of energy-conservation measures relating to exergy analysis. It should be pointed out that these diagrams have many uses in presenting the results of exergy analysis, but since most industrial energy systems are complex with many components and processes, the tabular form shown in Table 4.2 is generally required to show all exergy losses. These systems are generally too complex and extensive to be shown on single charts or diagrams, although individual processes may be diagrammed if it is desired to show results of exergy analysis for specific areas of a system.

CHAPTER 5

Examples of Exergy Analysis

The exergy analysis of a system starts with a flowchart of the energy flow through the system, including all the processes that affect a change in the energy level or the conditions of the working substance. The flowchart has interfaces with the system boundaries that for conventional terrestrial systems is the surrounding environment, plus any interfaces with other systems. The input energy requires interfacing with the fuel supply; air for combustion; and electrical power to motors that drive compressors, pumps, and fans. Output-energy interfaces involve the flue gas to the atmosphere, air from air conditioners, condenser heat to cooling water or cooling air, useful output from an electric generator or shaft, and waste heat that may be recovered as useful work or heat for another process or system. Interfaces are also required to other closed systems where a direct interaction takes place such as exists in combined-cycle power plants and staged refrigeration systems. In general, the new design of a combined-cycle power plant would be handled by a single exergy flowchart with both systems combined together so that parametric variations of the variables can be examined throughout the complete system.

The examples given in this chapter are selected to show the additional information provided by the exergy analysis over the conventional heat-balance analysis. Examples are given for several different types of energy-related system to show the versatility of the exergy-analysis approach.

5-1 INDIVIDUAL PROCESS BLOCKS

The block method of performing an exergy analysis of a system (described in Chapter 4) makes use of calculation blocks of processes organized into a flow-chart of the system. The equations used in calculating the exergy change, the

102

useful work, and the exergy loss for many energy-related processes were given in Chapter 4. To clarify the use of these equations, examples of these process calculations are given in this section. Each process is an independent set of calculations based on input data and assumed output requirements where they are specified. In each of these examples it is assumed that the process is in a system that has a surrounding environment of 70 °F temperature, 1-atm pressure, and a sea-level elevation. Each process is assumed to be adiabatic except where heat rejection or heat supply and work in or out of the system are specifically noted.

These process examples have the same block number as the equation blocks in Section 4-2. A list of the example blocks is given in Table 4-1, and the blocks are given in Appendix A.

5-2 HEAT AND POWER SYSTEMS

Applications of the exergy analysis in the literature have included several examples of fossil-fuel power plants (Bosnjakovic 1965; Baehr 1962). The power-plant example is effective in demonstrating the advantages of the exergy method since there are large irreversible losses in the combustion and heat-transfer processes that are not accounted for in the conventional heat balance. Furthermore, since power plants are the most energy-intensive systems in our industry, they have great potential for energy conservation through cycle improvements such as topping or bottoming loops.

Example 5-1. An example of a closed-cycle steam power plant is presented in three different ways to clearly show the exergy-analysis technique. First, a long narative form is used that discusses each process and calculation in detail. Then the analysis is made in the block form, and finally the calculations alone are summarized in tabular form. The results of these calculations are tabulated in the recommended exergy table form that shows the location, type, and magnitude of all exergy losses in the system.

Since this example is for a closed cycle, the exergy efficiency

$$\eta_{ex} = \frac{\text{Maximum exergy} - \Sigma \text{ exergy loss}}{\text{Maximum exergy}} = \frac{12,600 - 8800}{12,600} = 0.30$$

is the same as the heat-balance efficiency

$$\eta_{thermal} = \frac{\text{Useful work}}{\text{Heat input}} = \frac{3800}{12,600} = 0.30$$

This example of the exergy method does show that the distribution of the energy losses in the system is quite different from that for the losses in the heat-balance case. The heat balance indicates that a large part of the initial heat in the fuel is rejected to the condenser cooling water relative to other losses. However, when the quality of the energy is taken into consideration by the exergy analysis, the loss in the condenser is very small relative to the losses in the combustion process and during heat transfer between the combustion gas and the steam. If the efficiency of this steam plant is to be significantly improved, the exergy analysis indicates that the effort should be directed to the high-temperature furnace area where the great loss in the available work of the fuel occurs.

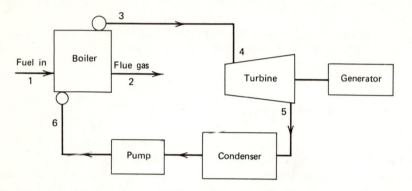

Figure 5-1. Power-plant cycle schematic.

This example is of a closed-cycle power plant shown in the schematic (Figure 5-1) and the temperature -entropy diagram, (Figure 5-2). As it is a closed-cycle system, the exergy efficiency and the conventional heat-balance efficiency will be the same. However, as is shown later, the exergy analysis will provide more detail and realistic performance data on the system and where the losses are in the system.

The assumptions made for this example are as follows:

Heating value of fuel	12,600 Btu/lb
Adiabatic combustion temperature	3500 °F
Furnace heat loss	12%
Flue gas:fuel mass flow ratio	16:1
Steam:fuel mass flow ratio	8:1
Steam pressure	1500 psia
Pressure drop, boiler to turbine	150 psia
Turbine efficiency	80%

The chemical energy in the fuel is considered as the initial value of the energy source since it can be theoretically recovered in devices such as a fuel cell with 100% efficiency. This value is considered to be the maximum exergy or work available from the fuel.

From the preceding assumptions, the operating conditions for each point in

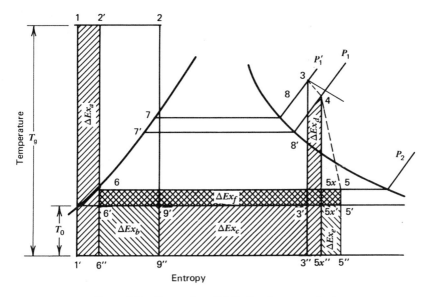

Figure 5-2. Exergy-loss distribution in power plant.

the system are shown in Table 5-1. These data were obtained from steam tables. The reference condition for the exergy calculations is that of water at 84 °F and, therefore, is the same as the feedwater to the boiler (point 6).

The heat given up by the combustion process represents the work capacity of the fuel and is shown in the $T-s$ diagram as a rectangle with a base of $1'-9''$ and a height T_g. This shows the change in entropy during combustion of the fuel.

Table 5-1 System Operating Conditions

Station	Temperature (°F)	Pressure (psia)	Enthalpy (Btu/lb)	Entropy (Btu/lb °F)	Exergy (Btu/lb fuel)
1	3500	14.7	12,600		12,600
2					11,100
3	900	1500	1,442	1.5713	4,880
4	895	1300	1,436	1.5778	4,800
5	84	0.6	960	1.770	190
6	84	0.6	52	0.10	4

Reference condition is liquid at 70 °F and atmospheric pressure. The effect of pressure is neglected since compressibility of the liquid is small; therefore, the properties are taken for liquid at 70 °F, which has a pressure of 0.36 psia. The enthalpy for the reference condition is 38.0 Btu/lb and the entropy, 0.0745 Btu/lb °R.

During combustion of 1 lb/hr of fuel, a certain steam flow rate is obtained on the basis of the properties and conditions of the steam in the boiler. The quantity of heat received by the steam is

$$Q_{steam} = m(h_3 - h_6)$$

Therefore, the steam flow for one lb/hr of fuel in this example is

$$m = \frac{(12,600)(1 - 0.12)}{1442 - 52} = 8.0 \, \text{lb/hr}$$

The change of state of the steam in the boiler is represented in the T-s diagram by 6–7–8–3. The vapor entering the turbine is represented by point 4. The steam entering the condenser is represented by point 5, and the constant temperature line 5–6 represents the condensation process.

Exergy Loss Resulting from Heat Loss during Combustion

This was assumed to be 12% and is considered as a direct exergy loss since it reduces the available work in the fuel by that amount.

$$Ex_a = (12,600)(0.12) = 1500 \, \text{Btu/lb}$$

This exergy loss is represented in the T-s diagram by the rectangle 1–$2'$–$6''$–$1'$–1.

Exergy Loss Resulting from Transfer of Chemical Energy to Heat Energy

During the combustion of the fuel, the remainder of the chemical energy

$$12,600 - 1500 = 11,100 \, \text{Btu/lb}$$

is transferred into thermal energy. This is represented by the area $2'$–2–$9''$–$6''$–$2'$ in the fuel and in the steam area 6–7–8–3–$3''$–$6''$–6.

During this the loss of exergy occurs because of two reasons. One reason is that the transfer of useful chemical energy into thermal energy is an irreversible process. The value of this area is

$$T_g - T_0 \frac{Q}{T_g} = Q \left[1 - \frac{T_0}{T_g} \right]$$

Therefore, the exergy loss is

$$Ex_b = m(h_3 - h_6) \frac{T_0}{T_g}$$

When the property values are entered into this equation, we obtain

$$Ex_b = 8.0 (1442 - 52) \frac{530}{3960} = 1490 \, \text{Btu/lb}$$

On the $T-s$ diagram this loss is represented by the area $6'-9'-9''-6''-6'$.

Exergy Loss Resulting from Heat Transfer between Gas and Steam

This is the second source of exergy loss in the process of heating the steam by combustion. The total exergy loss in this process can be determined from

$$T_0 \, \Delta s$$

where the change in entropy is shown in the $T-s$ diagram by the length of the line $6''-3''$. If this value is calculated and the value of Ex_b subtracted from it, the result will be Ex_c or the exergy loss in the transfer of heat between the gas and the steam. However, Ex_c can be directly calculated by the combined equation

$$Ex_c = T_0(m) \left[(s_3 - s_6) - \frac{h_3 - h_6}{T_g} \right]$$

Substituting the property values into the preceding equation gives

$$Ex_c = 530(8.0) \left[(1.5713 - 0.10) - \frac{1442 - 52}{3960} \right]$$

$$Ex_c = 4750 \, \text{Btu/lb}$$

This exergy loss is shown on the $T-s$ diagram by the rectangular area $9'-3'-3''-9'$. The total loss of exergy from the transformation of chemical energy into heat and from the nonuniform heat transfer, $Ex_b + Ex_c$, does not depend on the warm-temperature source under which the transformation and transfer occur. This total loss is shown on the $T-s$ diagram by the area $6'-3'-3''-6''-6'$, which is not connected with T_g in any way.

Exergy Loss during Steam Flow in Pipe from Boiler to Turbine

In this process there are two sources of exergy loss, the throttling of the steam flow resulting from friction and the nonequilibrium heat exchange between the steam and the surrounding medium. The total loss of exergy can be determined as the difference in the exergy at the beginning and the end of the process. The steam under condition 3 has the work capacity shown by the area $3-3'-6'-6-7-8-3$ and for 1 lb/hr of steam

$$Ex_3 = h_3 - h_6 - T_0(s_3 - s_6)$$

The steam in condition 4 has the work capacity shown by the area $4-5x'-6'-6-7'-8'-4$ and for 1 lb/hr of steam

$$Ex_4 = h_4 - h_6 - T_0(s_4 - s_6)$$

The exergy loss in the process is

$$Ex_d \quad Ex_3 - Ex_4$$

Substituting property values into the preceding equations, we obtain

$$Ex_3 = (1442 - 38) - 530(1.5713 - 0.0745)$$

$$= 611 \text{ Btu/lb steam}$$

$$= 4880 \text{ Btu/lb fuel}$$

$$Ex_4 = (1436 - 38) - 530(1.5778 - 0.0745)$$

$$= 601 \text{ Btu/lb steam}$$

$$= 4800 \text{ Btu/lb fuel}$$

$$Ex_d = Ex_4 - Ex_3 = 4880 - 4800 = 80 \text{ Btu/lb fuel}$$

Actually, the system here consists of steam and the surrounding medium, which serves as the cold sink. The entropy of 1 lb/hr of steam increases by $s_4 - s_3$. The medium receives $(h_3 - h_4)$ Btu/hr when T_0 = constant, and its entropy for 1 lb/hr of steam increases by $(h_3 - h_4)/T_0$. The overall increase in the entropy of the system for 1 lb/hr of steam is

$$s_{\text{total}} = (s_4 - s_3) + \frac{h_3 - h_4}{T_0}$$

and the loss of exergy for 1 lb/hr of steam is

$$Ex_d = T_0 \Delta s_{\text{total}} = T_0(s_4 - s_3) + (h_3 - h_4)$$

which will give the same value for Ex_d as previously obtained.

Exergy Loss during Irreversible Expansion of Steam in Turbine

This loss is determined from the basic equation

$$Ex_e = (m)T_0 \Delta s_{\text{total}}$$

and is calculated as follows

$$Ex_e = (m)T_0 \Delta s_{\text{total}}$$

$$= 8.0 \times 530(1.77 - 1.5778)$$

$$= 815 \text{ Btu/lb fuel}$$

In the $T-s$ diagram this loss is represented by the area $5x'-5'-5''-5x''-5x'$.

Exergy Loss from Nonuniform Heat Transfer in Condenser

This loss is determined from the basic equation

$$Ex_f = T_0\,\Delta s$$

In this case the change in entropy is

$$\Delta s = s_5 - s_{5x}$$

$$\Delta s = m\left[\frac{h_5 - h_6}{T_0}(s_5 - s_6)\right]$$

and the loss of exergy is

$$Ex_f = m\left[(h_5 - h_6) - T_0(s_5 - s_6)\right]$$

or by substituting in the property values

$$Ex_f = 8.0\left[(960 - 52) - 530(1.77 - 0.10)\right] = 186\;\text{Btu/lb fuel}$$

In the $T-s$ diagram this exergy loss is represented by the area $6-6-5'-6'-6$ or the area $6-5-5''-6''-6$ minus the area $6'-5'-5''-6''-6'$.

Useful Work of Steam in Turbine

$$\text{Work} = m(h_4 - h_5)$$
$$\text{Work} = 8.0(1436 - 960)$$
$$= 3800\;\text{Btu/lb fuel}$$

A summary of the preceding calculations is given in Table 5-2. The calculations are also shown in the exergy block form in Blocks 5-1–5-5. The results are shown in the standard exergy form in Table 5-3.

The conventional heat-balance analysis is performed as follows.

Heat loss during combustion $= Q(1-e) = 12,600(1 - 0.12)$
$= 1500\;\text{Btu/lb fuel}$

Heat loss in steam pipe $= m(h_3 - h_4) = 8.0(1442 - 1436)$
$= 48\;\text{Btu/lb fuel}$

Heat transferred to useful work $= m(h_4 - h_5) = 8.0(1436 - 960)$
$= 3800\;\text{Btu/lb fuel}$

Heat given up in condenser $= m(h_5 - h_6) = 8.0(960 - 52)$
$= 7260\;\text{Btu/lb fuel}$

Table 5-2   Exergy-Analysis Calculations

Combustion heat release: 12,500 Btu/lb
Adiabatic gas temperature: 3960 °R
Combustion efficiency: 88%   (analysis based on per lb fuel)

Ex_1 = 12,600 Btu/lb
Loss resulting from furnace inefficiency
$\quad Ex_a$ = 12,600(1 − 0.88) = 1500 Btu/lb
Ex_2 = 12,600 − 1500 = 11,100 Btu/lb
Heat introduced into steam
$\quad h_3 - h_6$ = 1442 − 52 = 1390 Btu/lb
Steam flow rate
$\quad$ 11,100/1390 = 8.0 lb steam/lb fuel
Total exergy loss fuel to steam
$\quad Ex_b + Ex_c$ = 8.0(530)(1.5713 − 0.100) = 240 Btu/lb
Exergy loss resulting from chemical-to-heat change
$\quad Ex_b$ = 11,100(530/3960) = 1490 Btu/lb
Exergy loss resulting from gas-to-steam heat transfer
$\quad Ex_c$ = 6240 − 1490 = 4750 Btu/lb
Ex_3 = 11,100 − 6240 = 4880 Btu/lb
Ex_4 = [(1436 − 38) − 530(1.5778 − 0.0745)](8.0) = 4800 Btu/lb
Exergy loss in piping
$\quad Ex_d$ = $Ex_4 - Ex_3$ = 80 Btu/lb
Exergy loss in turbine
$\quad Ex_e$ = 8.0(530)(1.770 − 1.5778) = 810 Btu/lb
Ex_5 = 8.0[(960 − 38) − 530(1.770 − 0.0745)] = 190 Btu/lb
Total exergy change in turbine
$\quad Ex_e$ + Work = $Ex_4 - Ex_5$ = 4800 − 190 = 4610 Btu/lb
Turbine work = 4610 − 810 = 3800 Btu/lb
Exergy loss from condenser
$\quad Ex_6$ = [(52 − 38) − 530(0.100 − 0.0745)]8 = 4 Btu/lb
$\quad Ex_f$ = $Ex_5 - Ex_6$ = 190 − 4 = 186 Btu/lb

These heat-balance results are also shown on the exergy-analysis form in Table 5-3 for comparison with the exergy-balance results.

Although the efficiencies of this closed cycle by the heat-balance and the exergy methods are the same, the results in Table 5-3 show that a clearer definition of the real losses in the system is given by the exergy method of analysis.

Simple Steam Power Plant (Units in Btu/lb Fuel)

Station or Component	Exergy	Change in Exergy	Heat Transfer	Useful Work	Exergy Losses						Total Exergy Loss (%)
					Chemical Change	Heat Transfer	Friction	Machine Inefficiency	Heat Rejection	Total	
1 Furnace	12,600	1500							1500	1500	17
2 Boiler	11,100	6240	11,100		1490	4750				6240	71
3 Piping	4,880	80	48			48	32			80	1
4 Turbine	4,800	4610		3800				810		810	9
5 Condenser	190	186							186	186	2
6	4									8800	100

Exergy Balance

Loss of exergy during combustion = Ex_a = 1500
Loss of exergy during irreversible transfer of chemical energy to heat = Ex_b = 1490
Loss of exergy during transfer of heat from gas to steam = Ex_c = 4750
Loss of exergy in piping = Ex_d = 80
Loss of exergy during irreversible expansion in turbine = Ex_e = 810
Loss of exergy through heat transfer in condenser = Ex_f = 186

Total exergy loss = 8800
12,600 − 8800 = 3800 useful work

Heat Balance

Loss of heat during combustion = 1500
Loss of heat in piping = 48
Heat rejected to coolant in condenser = 7260
Heat transferred into useful work = 3800

Block 5-1

Process: Combustion (process block 1–2)

Component: Steam-generator furnace

Combustion heat release = 12,600 Btu/lb fuel
Furnace efficiency   = 88% (because of incomplete combustion)
Reference temperature   = 530 °R
Combustion temperature   = 3960 °R

Ex_1 = 12,600 Btu/lb fuel	Point 1

Loss resulting from furnace inefficiency

Ex_{loss} = 12,600(1 − 0.88) = 1500 Btu/lb
Ex_{1a} = 12,600 − 1500 = 11,100 Btu/lb

Combustion loss (chemical-to-heat conversion)

Ex_{loss} = $Q(T_0/T)$ = 11,100(530/3960) = 1490 Btu/lb

Ex_2 = Ex_1 − Ex_{losses}
Ex_2 = 12,600 − 1500 − 1490 = 9610 Btu/lb

Ex_2 = 9610 Btu/lb fuel	Point 2

Block 5-2

Process: Heat transfer (process block 6–3)

Component: Steam-generator boiler

Final Steam Conditions	Initial Steam Conditions	Reference Conditions
h_3 = 1442 Btu/lb steam	h_6 = 52 Btu/lb steam	h_0 = 38 Btu/lb steam
s_3 = 1.5713 Btu/lb °R	s_6 = 0.100 Btu/lb °R steam	s_0 = 0.0745 Btu/lb °R steam

Flow rate: 8 lb/hr steam · 1 lb/hr fuel
Combustion gas: Ex_2 = 9610 Btu/lb fuel

Steam Ex_6 = (52 − 38) − 530(0.100 − 0.0745) = 4 Btu/lb	Point 6

Exergy loss resulting from heat transfer

$$Ex_{loss} = mT_0[(s_3 - s_6) - Q/T_g] \text{ where } \frac{Q}{T_g} = \frac{(h_3 - h_6)}{T_g}$$

Total exergy loss	Loss in chemical-to-heat conversion

$$Ex_{loss} = (8)530[(1.5713 - 0.10) - (1442 - 52)/3960]$$
$$= 4750 \text{ Btu/lb fuel}$$

$$Ex_3 = (1442 - 38) - 530(1.5713 - 0.0745)$$
$$= 610 \text{ Btu/lb steam} = 4880 \text{ Btu/lb fuel}$$

Point 3

Block 5-3

Process: Friction (process block 3–4)

Component: Piping, boiler – turbine

Initial Steam Conditions	Final Steam Conditions	Reference Conditions
$h_3 = 1442 \text{ Btu/lb}$	$h_4 = 1436 \text{ Btu/lb}$	$h_0 = 38 \text{ Btu/lb}$
$s_3 = 1.5713 \text{ Btu/lb}^\circ R$	$s_4 = 1.5778 \text{ Btu/lb}^\circ R$	$s_0 = 0.0745 \text{ Btu/lb}^\circ R$

$$Ex_3 = 8[(1442 - 38) - 530(1.5713 - 0.0745)] = 4880 \text{ Btu/lb fuel} \quad \text{Point } 3$$

$$Ex_{loss} = mT_0[(s_4 - s_3) - (Q/T_g)]$$
$$Ex_{loss} = (8)530[(1.5778 - 1.5713) - (1336 - 1442)/900]$$
$$= 80 \text{ Btu/lb fuel}$$

Numerical values are rounded off

$$Ex_4 = 8[(1436 - 38) - 530(1.5778 - 0.0745)] = 4800 \text{ Btu/lb fuel} \quad \text{Point } 4$$

Block 5-4

Process: Expansion with work (process block 4–5)

Component: Steam turbine

Initial Steam Conditions	Final Steam Conditions	Reference Conditions
$h_4 = 1436$ Btu/lb	$h_5 = 960$ Btu/lb	$h_0 = 38$ Btu/lb
$s_4 = 1.5778$ Btu/lb $^\circ$R	$s_5 = 1.770$ Btu/lb $^\circ$R	$s_0 = 0.0745$ Btu/lb $^\circ$R
Turbine efficiency: 80%		

$$Ex_4 = 8[(1436 - 38) - 530(1.5778 - 0.0745)] = 4800 \text{ Btu/lb fuel} \qquad \text{Point } 4$$

Total exergy change

$$Ex_5 - Ex_4 = 4800 - 190 = 4610 \text{ Btu/lb fuel}$$

Total useful work

$$W = (8)(1436 - 960) = 3800 \text{ Btu/lb fuel}$$

Exergy loss resulting from irreversibility

$$Ex_{loss} = (8)530(1.770 - 1.5778) = 815 \text{ Btu/lb fuel}$$

$$Ex_5 = 8[(960 - 38) - 530(1.770 - 0.0745)] = 190 \text{ Btu/lb fuel} \qquad \text{Point } 5$$

Block 5-5

Process: Condensing and heat rejection (process block 5–6)

Component: Condenser

Initial Steam Conditions	Final Condensate Conditions	Reference Conditions
$h_5 = 960$ Btu/lb	$h_6 = 52$ Btu/lb	$h_0 = 38$ Btu/lb
$s_5 = 1.770$ Btu/lb $^\circ$R	$s_6 = 0.100$ Btu/lb $^\circ$R	$s_0 = 0.0745$ Btu/lb $^\circ$R

$$Ex_5 = 8[(960 - 530(1.77 - 0.0745)] = 190 \text{ Btu/lb fuel} \qquad \text{Point } 5$$

Total exergy change in condenser is

$$Ex_5 - Ex_6 = 190 - 0 = 190 \text{ Btu/lb fuel}$$

If none of the heat is recovered as work, total exergy loss is

$$Ex_{loss} = (8)[530(1.770 - 0.10) - (960 - 52)] = 190 \text{ Btu/lb fuel}$$

$Ex_6 = 8[(52 - 38) - 530(0.100 - 0.0745)] = 4 \text{ Btu/lb fuel}$	Point 6

Example 5-2. This example of a nuclear power plant with single and dual gas cycles shows the difference in efficiency for systems that are other than single, closed cycles. The improvement gained by adding a bottoming loop is shown by combining an isobutane cycle with the basic helium cycle. The exergy analysis accounts for the irreversible heat transfer in the helium to isobutane heat exchanger, which is not done in the conventional heat-balance analysis.

This example is of a nuclear power plant in both single- and a dual-cycle configurations. The single cycle has a helium Brayton cycle, whereas the dual system has an isobutane Rankine cycle as a bottoming unit to the helium Brayton cycle. The schematic for the single cycle is shown in Figure 5-3 and for the dual cycle, in Figure 5-4. The enthalpy–entropy diagram for the helium cycle is shown in Figure 5-5, and Figure 5-6 gives the pressure–enthalpy diagram for the isobutane Rankine cycle.

The system operating conditions are given in Table 5-4. The detailed calculations of the exergy values in Table 5-4 are provided as follows.

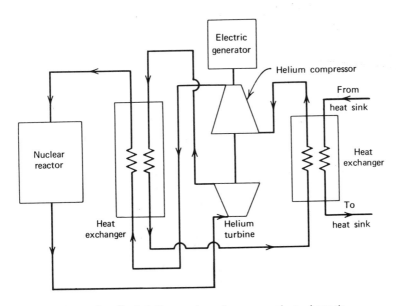

Figure 5-3. Single helium-cycle nuclear power plant schematic.

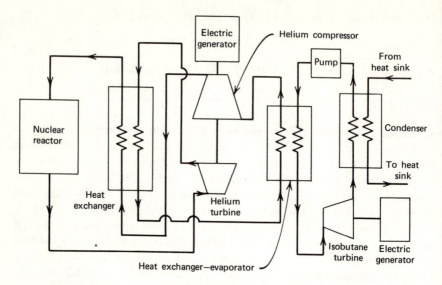

Figure 5-4. Dual helium–isobutane-cycle nuclear power plant schematic.

Table 5-4   System Operating Conditions

System	Temperature (°F)	Pressure (psia)	Enthalpy (Btu/lb)	Entropy (Btu/lb °F)	Exergy (Btu/lb helium)
Helium					
1	140	425	755	5.980	444
2	335	975	1194	6.100	820
3	920	975	1683	6.575	1057
4	1300	975	2318	6.996	1469
5	1000	425	1830	7.096	928
6	425	425	1097	6.451	537
1	140	425	755	5.980	444
Isobutane					
7	90	45	40	0.8928	0
8	95	1000	56	0.915	7
9	375	1000	270	1.220	94
10	165	45	226	1.232	10
7	90	45	40	0.8928	0

116

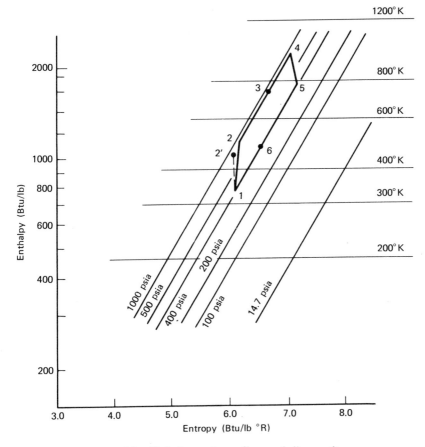

Figure 5-5.   Enthalpy–entropy diagram, helium cycle.

The reference conditions for exergy calculations are as follows:

	Helium	Isobutane
T_0	$530\,^{\circ}R$	$530\,^{\circ}R$
h_0	$162.7\,Btu/lb$	$40\,Btu/lb$
s_0	$5.701\,Btu/lb\,^{\circ}R$	$0.8928\,Btu/lb\,^{\circ}R$

The exergy entering the compressor is

$$Ex_1 = (755 - 162.7) - 530(5.98 - 5.701)$$

$$= 444.43\,Btu/lb$$

118

Examples of Exergy Analysis

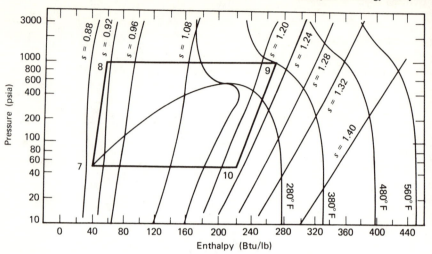

Figure 5-6. Pressure-enthalpy diagram, isobutane.

In the compressor the exergy of the helium is increased because of the potential for work put into it by its increase in pressure. Under ideal compression (reversible) the exergy after compression from 420 psia to 973 psia would be

$$Ex_2' = (1071 - 162.7) - 530(5.98 - 5.701)$$

$$= 760.43 \text{ Btu/lb}$$

However, since the compressor is only 72% efficient, the actual exergy added to the helium to achieve the desired pressure is

$$Ex_2 = (1193.9 - 162.7) - 530(6.100 - 5.701)$$

$$= 819.7 \text{ Btu/lb}$$

In the process of compression the exergy introduced by the shaft is not all converted to exergy in the helium since some is lost through irreversibility in the compressor. The exergy loss in the process is the difference between the exergy of the ideal reversible and the actual process,

Exergy loss in the compressor ΔEx_{1-2} = 819.70 − 760.4

$$= 59.38 \text{ Btu/lb}$$

The helium gas leaving the compressor passes through a recuperative heat exchanger, where it is heated by the gas that is leaving the turbine. In a heat exchanger the heat energy given up by one fluid stream ($Q_1 = w_1 c_{p_1} \Delta T_1$) is completely transferred to the other fluid stream ($Q_2 = w_2 c_{p_2} \Delta T_2$) except for heat loss through heat leak to or from the heat exchanger. This heat leak is usually negligible for well-insulated systems, and it is considered as negligible in this example. The heat transfer in the heat exchanger is accompanied by a loss of exergy as a result of the irreversibilities in the heat-exchange process. The exergy change in the low-temperature fluid in the heat exchanger is considered at this time, whereas the exergy change in the high-temperature fluid along with the exergy loss is considered later in the analysis of the system.

The exergy entering the recuperative heat exchanger is the exergy leaving the compressor, 819.7 Btu/lb, since we assume that there is no exergy loss in the piping.

The exergy of the low-temperature stream leaving the recuperative heat exchanger is

$$Ex_3 = (1683 - 162.7) - 530(6.5747 - 5.701)$$

$$= 1057.24 \text{ Btu/lb}$$

This is also the exergy of the fluid stream entering the reactor.

In the reactor exergy is added to the helium stream in the form of heat from the reactor-fuel elements. The temperature of the fuel element is limited by the material and in this case is assumed to be 1800 °F. The exergy loss in transferring the heat from the fuel element to the helium gas is

$$\text{Heat transferred } Q = 2318 - 1683 = 635 \text{ Btu/lb}$$

$$Ex = 635 \, \frac{530\,°\text{R}}{1960\,°\text{R}} = 172 \text{ Btu/lb}$$

The exergy leaving the reactor is

$$Ex_4 = (2318 - 162.7) - 530(6.9964 - 5.701)$$

$$Ex_4 = 1468.74 \text{ Btu/lb}$$

After leaving the reactor, the helium flows to the turbine. There are frictional losses in this flow process, as shown in Example 5-1, which are neglected here to simplify this example of a dual-cycle system. The exergy leaving the turbine under ideal reversible conditions is

$$Ex_{5'} = (1750 - 162.7) - 530(6.9964 - 5.701)$$

$$= 900.74 \text{ Btu/lb}$$

However, because of the inefficiency in the turbine expansion process, exergy losses are incurred and the actual exergy leaving the turbine is

$$Ex_5 = (1830 - 162.7) - 530(7.0957 - 5.701)$$

$$= 928.11 \text{ Btu/lb}$$

The exergy loss in the turbine through irreversibility is

$$Ex_{4-5} = 928.11 - 900.74 = 27.37 \text{ Btu/lb}$$

From the turbine the helium flows to the recuperative heat exchanger, where exergy is given up to the cold helium stream previously calculated. The exergy of the warm helium gas leaving the recuperative heat exchanger is

$$Ex_6 = (1097 - 162.7) - 530(6.4506 - 5.701)$$
$$= 537.01 \text{ Btu/lb}$$

The exergy given up by the warm helium gas stream in the recuperative heat exchanger is

$$928.11 - 537.01 = 391.90 \text{ Btu/lb}$$

The exergy taken up by the cold helium gas stream as calculated previously for Ex_2 and Ex_3 is

$$1057.24 - 819.0 = 238.0 \text{ Btu/lb}$$

The difference between the exergy given up by the warm helium stream and the exergy absorbed by the cold helium gas stream is

$$391.90 - 238.0 = 154.0 \text{ Btu/lb}$$

and this is the exergy loss through the irreversibility of heat transfer in the heat exchanger. As discussed in Chapter 4, the magnitude of this exergy loss is a function of the mean temperature difference between the two fluid streams.

The difference between the exergy of the warm helium gas leaving the recuperator and the exergy of the gas entering the compressor is the exergy given up to the secondary fluid in the dual cycle

$$Ex_6 - Ex_1 = 537.01 - 444.43 = 93.58 \text{ Btu/lb helium}$$

Assuming no heat leak to or from the heat exchanger that transfers heat from the helium to the secondary fluid, isobutane, for the isobutane flow rate we have

$$\frac{h_6 - h_1}{h_9 - h_8} = \frac{1097 - 755}{270 - 56} = 1.60 \text{ lb isobutane/lb helium}$$

To keep all values of exergy in the system on a common basis, they are specified on the basis of one pound of helium flow per unit time.

The isobutane cycle operates within 45−1000 psia, as shown in the pressure−enthalpy diagram (Figure 5-6). The enthalpy scale is based on a zero reference of saturated liquid at $0\,^\circ\text{F}$. The reference data for the exergy calculations are

$$T_0 = 530\,^\circ\text{R} \,(70\,^\circ\text{F})$$
$$h_0 = 40 \text{ Btu/lb}$$
$$s_0 = 0.8928 \text{ Btu/lb}\,^\circ\text{R}$$

The exergy entering the isobutane pump is

$$Ex_7 = (40 - 40) - 530(0.8928 - 0.8928)$$

$$Ex_7 = 0 \text{ Btu/lb}$$

The isobutane is pumped to 1000 psia, where under ideal isentropic conditions of compression the exergy would be

$$Ex_8' = (44 - 40) - 530(0.8928 - 0.8928)$$

$$= 4 \text{ Btu/lb isobutane}$$

$$= 6.4 \text{ Btu/lb helium}$$

But since the pump is only 80% efficient the actual value of the exergy after pumping to 1000 psia is

$$Ex_8 = (56 - 40) - 530(0.915 - 0.8928)$$

$$Ex_8 = 4.23 \text{ Btu/lb isobutane}$$

$$= 6.77 \text{ Btu/lb helium}$$

The isobutane flows through the heat exchanger in which heat is transferred to the isobutane from the helium. The heat added raises the isobutane enthalpy to 270 Btu/lb. At this point the entropy is 1.22 Btu/lb $^{\circ}$R, and the temperature is 375 $^{\circ}$F. The exergy at this point is

$$Ex_9 = (270 - 40) - 530(1.22 - 0.8928)$$

$$Ex_9 = 56.58 \text{ Btu/lb isobutane}$$

$$Ex_9 = 93.9 \text{ Btu/lb helium}$$

The difference between the exergy at point 9 and the exergy at point 8 is the exergy picked up by the isobutane from the helium, 87 Btu/lb helium. The exergy given up by the helium calculated before is 93 Btu/lb helium, and the difference is the loss in the heat exchanger encountered through the irreversibility during the heat transfer.

From the heat exchanger the isobutane goes to the turbine, where it expands from 1000 psia to 45 psia, producing work. Under an ideal isentropic expansion the exergy at the turbine outlet would be

$$Ex_{10}' = (221 - 40) - 530(122 - 0.8928)$$

$$= 758 \text{ Btu/lb isobutane}$$

$$= 12.6 \text{ Btu/lb helium}$$

However, with a turbine efficiency of 90% the actual exergy at the turbine outlet will be

Table 5-5 Exergy Chart for Single Helium System

Single Helium Cycle (All Values per lb/hr Helium Flow)

Station or Component	Exergy	Change in Exergy	Heat Transfer	Useful Work	Chemical Change	Exergy losses Heat Transfer	Friction	Machine Inefficiency	Heat Rejection	Total	Total Exergy Loss (%)
1	444										
Compressor		376		−316				60		60	12
2	820										
Heat exchanger		237	489			77				77	15
3	1057										
Reactor		412	635			172				172	35
4	1469										
Turbine		−541		514				27		27	5
5	928										
Heat exchanger		−391	733			77				77	15
6	537										
Heat exchanger		−93							93	93	18
1	444										
Totals				198						506	100

Dual Helium–Isobutane Cycle

(Isobutane: Helium Flow Rate, 1.6:1)
(All Values per lb/hr Helium Flow)

| Station or Component | Exergy | Change in Exergy | Heat Transfer | Useful Work | Chemical Change | Exergy Losses | | | | | Total |
						Heat Transfer	Friction	Machine Inefficiency	Heat Rejection	Total	Total Exergy Loss (%)
1	444										
Compressor 2	820	376		−316				60		60	14
Heat exchanger 3	1057	237	489			77				77	18
Reactor 4	1469	412	635			172				172	40
Turbine 5	928	−541		514				27		27	6
Heat exchanger 6	537	−391	733			77				77	18
Heat exchanger 1		−93				3				3	0.5
7	0										
Compressor 8	7	7		−6				1		1	0
Heat exchanger 9	94	87				3				3	0.5
Turbine 10	10	−84		81				3		3	1
Compressor 7	0	−10							10	10	2
Totals				273						433	100

$$Ex_{10} = (226 - 40) - 530(1.232 - 0.8928)$$

$$Ex_{10} = 6.12 \text{ Btu/lb isobutane}$$

$$Ex_{10} = 10.2 \text{ Btu/lb helium}$$

The difference between Ex'_{10} and Ex_{10} is the exergy loss through irreversibility in the turbine expansion process. This exergy loss is

$$Ex'_{10} - Ex_{10} = 12.6 - 10.2 = 2.4 \text{ Btu/lb helium}$$

The work performed in the turbine is then

$$93.9 - 12.6 = 81.3 \text{ Btu/lb helium}$$

The isobutane from the turbine goes to the condenser, where it is condensed to the condition entering the pump. The exergy entering the condenser is 10.2 Btu/lb helium, whereas the exergy leaving the condenser and entering the pump is zero, according to our referenced exergy value. Therefore, the exergy given up by the isobutane in the condensation process is 10.2 Btu/lb helium. This exergy is considered lost exergy since it is not recovered.

The results of the single-cycle calculations are given in Table 5-5 and those for the dual cycle, in Table 5-6. A comparison of the heat-balance and exergy efficiencies is shown in Table 5-7. Although the heat-balance efficiencies are

Table 5-7   Comparison of Energy and Exergy Efficiencies

Heat-Balance Efficiency	Exergy Efficiency
Single helium cycle	
Input	Input
$\quad$ 2318 − 1683 = $\quad$ 635	$\quad$ 1469 − 1057 = 412
$\quad$ 1194 − $\quad$755 = $\quad$ 439	$\quad$ 820 − $\quad$444 = 376
$\quad\quad\quad\quad\quad\quad\quad\quad$ 1074 Btu/lb total	$\quad\quad\quad\quad\quad\quad\quad$ 788 Btu/lb total
Output	Exergy losses
$\quad$ 2318 − 1830 = 488 Btu/lb	$\quad$ 60 + 172 + 27 + 93 + 154 = 506
$\eta = (488/1074) = 0.45$	$\eta = [(788 - 506)/788] = 0.36$
Dual helium−isobutane cycle	
Input	Exergy input
$\quad$ 635 + 439 = 1074	$\quad$ 412 + 376 = 788
$\quad\quad$ 56 − $\quad$40 = $\quad\quad$ 16	$\quad\quad$ 7 − $\quad\quad$0 = $\quad\quad$ 7
$\quad\quad\quad\quad\quad\quad\quad\quad$ 1090 Btu/lb total	$\quad\quad\quad\quad\quad\quad\quad$ 795 Btu/lb total
Output	Exergy losses
$\quad$ 2318 − 1830 = 488	$\quad$ 60 + 172 + 27 + 93 + 154 + 10
	$\quad\quad\quad\quad$ + 347 = 433
$\quad\quad$ 270 − $\quad$226 = $\quad$ 44	
$\quad\quad\quad\quad\quad\quad\quad\quad$ 532 Btu/lb total	
$\eta = (532/1090) = 0.49$	$\eta = [(795 - 433)/795] = 0.46$

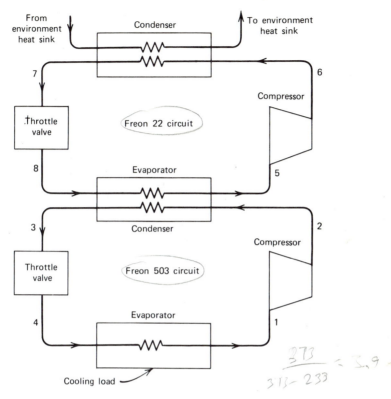

Figure 5-7.   Two-stage refrigeration system schematic.

higher for the two cycles, the maximum gain in adding the secondary cycle is for the exergy efficiency.

5-3   REFRIGERATION SYSTEMS

Because of the low Carnot efficiencies of refrigeration systems, especially in the low cryogenic temperature range, the application of exergy analysis to these systems can provide information with which effective energy-conservation gains can be made. The examples given here cover conventional air-conditioning systems as well as low-temperature cryogenic systems.

The significant effect of irreversible losses in cryogenic systems can be seen in the examples, and the influence on power requirements relative to the useful work produced is readily noted.

The advantage of using the exergy method of analysis for air-conditioning

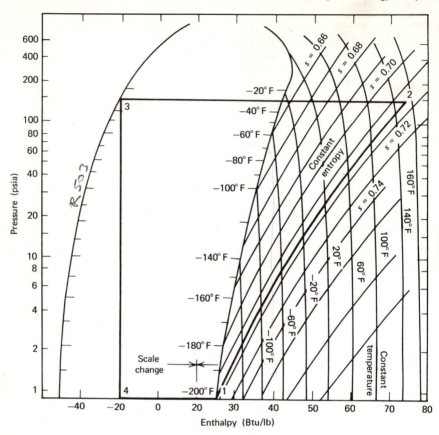

Figure 5-8. *P–h* Diagram for Freon 503 cycle.

systems is not so apparent unless the high-quality energy of the fossil fuel used to generate the electrical power for the air conditioner is considered. In this case a mismatch exists between the high-quality fossil-fuel or nuclear power-plant energy and the low-quality heat transferred in the air-conditioning process. A better match of energy levels is made if the air conditioner is designed to operate on low-quality energy such as waste heat or solar panel heat. In this system the irreversible losses in the total system are small, but they are very significant relative to the low-quality available energy. If solar heating is to be effectively used for domestic air conditioning, it is necessary to achieve the highest possible thermodynamic efficiency by practical elimination of all system losses associated with the irreversible production of entropy.

Example 5-3. This is an example of a two-stage refrigeration system using Freon 22 and Freon 503. A baseline case is given in which the compressor

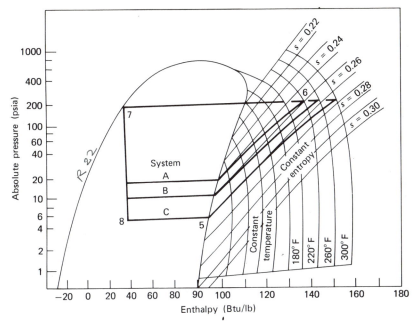

Figure 5-9. $P-h$ Diagram for Freon 22 cycle.

efficiencies are assumed to be 100%. Three variations of this system are then analyzed with an 80% compressor efficiency and with the Freon 22 evaporator pressure having values of 15 psia, 10 psia and 5 psia. These different evaporator pressures result in a variation of the temperature difference between the two refrigerants in the interconnecting heat exchanger.

A schematic of the system is shown in Figure 5-7. Pressure–enthalpy diagrams are given in Figure 5-8 for the Freon 503 cycle and in Figure 5-9 for the Freon 22 cycle. The calculated data are presented in Table 5-8 for both the Freon 503 and Freon 22 circuits. Tables 5-9–5-12 are the exergy analysis sheets for the four different systems. Exergy losses are shown in these sheets. Table 5-13 presents a comparison of the energy and exergy balances.

Example 5-4.   A closed-cycle, reverse Brayton cryogenic cooler that provides 100 W of refrigeration at 30 °R is analyzed. The basic performance of the system is taken from NBS TN-366 by Muhlanhaupt and Strobridge (1968).

The cycle (Figure 5-10) uses helium as the working fluid, and refrigeration is accomplished by a turboexpander. A recuperative heat exchange is used to recover the refrigeration available in the gas returning to the compressor. In this example the temperature difference in the recuperative heat exchanger is varied to show the influence on the system performance of the irreversible heat transfer.

Table 5-8 System Performance Data (Units in lb Freon 503)

Station	Freon 503				Freon 22			
	1	2	3	4	5	6	7	8
Reference	$T_0 = 530°R, h_0 = 60.0, s_0 = 0.7450$				$T_0 = 530°R, h_0 = 117, s_0 = 0.2760$			
				Baseline				
Pressure (psia)	0.88	150	150	0.88	15	200	200	15
Temperature (°F)	−200	153	−30	−200	−40	180	96	−40
Enthalpy (Btu/lb)	25	74	−20	−20	100	129	38	38
Entropy (Btu/lb °R)	0.714	0.714	0.511	0.543	0.239	0.239	0.077	0.091
Exergy (Btu/lb)	−18.57	30.43	44.02	27.06	3.97	48.05	40.3	28.96
				System A				
Pressure (psia)	0.88	150	150	0.88	15	200	200	15
Temperature (°F)	−200	153	−30	−200	−40	215	96	40
Enthalpy (Btu/lb)	25	74	−20	−20	100	136	38	38
Entropy (Btu/lb °R)	0.714	0.714	0.511	0.543	0.239	0.249	0.077	0.091
Exergy (Btu/lb)	−18.57	30.43	44.02	27.06	3.97	56.96	45.3	32.58
				System B				
Pressure (psia)	0.88	150	150	0.88	10	200	200	10
Temperature (°F)	−200	153	−30	−200	−56	240	96	−56
Enthalpy (Btu/lb)	25	74	−20	−20	98	141	38	38
Entropy (Btu/lb °R)	0.714	0.714	0.511	0.543	0.243	0.257	0.077	0.087
Exergy (Btu/lb)	−18.57	30.43	44.02	27.06	2.67	60.30	46.9	44.02
				System C				
Pressure (psia)	0.88	150	150	0.88	5	200	200	5
Temperature (°F)	−200	153	−30	−200	−79	278	96	−79
Enthalpy (Btu/lb)	25	74	−20	−20	96	148	38	38
Entropy (Btu/lb °R)	0.714	0.714	0.511	0.543	0.253	0.267	0.077	0.081
Exergy (Btu/lb)	−18.57	30.43	44.02	27.06	−16.12	65.46	48.5	44.56

EXERGY LOSSES

Table 5-9 Baseline System (Units in Btu/lb Freon 503)

Flow Rate, Freon 22/Freon 503 = 1.52

Station or Component	Exergy	Change in Exergy	Heat Transfer	Work	Chemical Change	Exergy Losses Heat Transfer	Friction	Machine Inefficiency	Heat Rejection	Total	Total Exergy Loss (%)
1	-18.6										
Compressor		49.0		-49.0							
2	30.4										
Condenser		13.6	-94.0			5.7				5.7	12
3	44.0										
Expansion		-17.0						17.0		17.0	36
4	27.0										
Evaporator		-45.6	45.0								
1	-18.6										
5	4.0										
Compressor		44.0		-44.0							
6	48.0										
Condenser		-7.7	-91						7.7	7.7	16
7	40.3										
Expansion		-11.3								11.3	24
8	29.0										
Evaporator		25.0	94.0			5.7				5.7	12
5	4.0									47.1	100

R503 CIRCUIT

R22 CIRCUIT

129

Table 5-10   System A: 15 psia Freon 22 Evaporator (Units in Btu/lb Freon 503)

Flow Rate, Freon 22/Freon 503 = 1.71

Station or Component	Exergy	Change in Exergy	Heat Transfer	Work	Exergy Losses						Total Exergy Loss (%)
					Chemical Change	Heat Transfer	Friction	Machine Inefficiency	Heat Rejection	Total	
1	−18.6										
Compressor		50.4		−38.9				10.6		10.6	15
2	31.8										
Condenser		12.0	−106			8.0				8.0	11
3	44.0										
Expansion		−17.0						17.0		17.0	23
4	27.0										
Evaporator		−45.6	45.0								
1	−18.6										
5	4.5										
Compressor		52.5		−47.2				5.3		5.3	7
6	57.0										
Condenser		11.7		−98					11.7	11.7	16
7	45.3										
Expansion		12.3						12.3		12.3	17
8	33.0										
Evaporator		−28.0				8.0				8	11
										72.9	100

Table 5-11 System B: 10 psia Freon 22 Evaporator (Units in Btu/lb Freon 503)

Flow Rate, Freon 22/Freon 503 = 1.77

Station or Component	Exergy	Change in Exergy	Heat Transfer	Useful Work	Chemical Change	Exergy Losses — Heat Transfer	Friction	Machine Inefficiency	Heat Rejection	Total	Total Exergy Loss (%)
1	−18.6										
Compressor		50.4		−39.8				10.6		10.6	12
2	31.8										
Condenser		12.0	−106			14.0				14	16
3	44.0										
Expansion		−17.0	45.0					17.0		17	20
4	27										
Evaporator		−45.6									
1	−18.6										
5	− 2.7										
Compressor		63.0		−55.6				7.4		7.4	9
6	60.3										
Condenser		13.4	−103						13.4	13.4	16
7	46.9										
Expansion		9.4						9.4		9.4	11
8	37.5										
Evaporator		−40.0				14.0				14	
5	− 2.7									85.8	100

131

Table 5-12 System C: 5 psia Freon 22 Evaporator (Units in Btu/lb Freon 503)

Flow Rate, Freon 22/Freon 503 = 1.83

Station or Component	Exergy	Change in Exergy	Heat Transfer	Useful Work	Exergy Losses						Total Exergy Loss (%)
					Chemical Change	Heat Transfer	Friction	Machine Inefficiency	Heat Rejection	Total	
1	−18.6										
Compressor		50.0		−39.8				10.6		10.6	11
2	31.8										
Condenser		12.0	−106			24.4				24.4	23
3	44.0										
Expansion		−17.0						17.0		17.0	16
4	27.0										
Evaporator		−45.6	45								
1	−18.6										
5	−16.1										
Compressor		81.6		−74.2				7.4		7.4	7
6	65.5										
Condenser		−17.0	−110						17	17	16
7	48.5										
Expansion		3.9						3.9		3.9	4
8	44.6										
Evaporator		−60.8				24.4				24.4	
										104.7	100

Table 5-13   Comparison of Energy and Exergy Balances

Heat-balance Efficiency	Exergy Efficiency		
	Baseline		
Input	Input exergy		
$(100) + (74 + 25) = 111\,\text{Btu/lb}$	503 compr 49		Exergy loss
Output	22 compr 44		47.1
45 Btu/lb	—		
	Total	93 Btu/lb	
$\eta = (45/111) = 0.41$	$\eta = (93 - 47.1)/93 = 0.49$		
	System A		
Input	Input exergy		
111 Btu/lb	503 compr 50.4		Exergy loss
	22 compr 52.5		72.9
	Total	102.9 Btu/lb	
$\eta = (45/111) = 0.41$	$\eta = (102.9 - 72.9)/102.9 = 0.29$		
	System B		
Input	Input exergy		Exergy loss
$(90 - 38) + (74 - 25) = 109$	503 compr 50.4		85.8
Output	22 compr 63		
45 Btu/lb	—		
	total	113.4 Btu/lb	
$\eta = (45/109) = 0.41$	$\eta = (113.4 - 85.8)/113.4 = 0.24$		
	System C		
Input	Input exergy		Exergy loss
$(96 - 38) + (74 - 25) = 107$	503 compr 50		104.7
Output	22 compr 82		
45 Btu/lb	—		
	Total	132 Btu/lb	
$\eta = (45/107) = 0.42$	$\eta = (132 - 104.7)/132 = 0.21$		

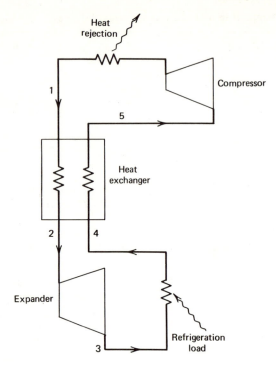

Figure 5-10. Schematic of reverse Brayton cycle.

For simplicity only the refrigeration part of the cycle, the expander, the recuperative heat exchanger, and the refrigeration load are considered in this analysis, with the compression taken as an ideal isothermal process (no irreversible losses). Table 5-14 lists the conditions assumed for the cycle and the exergy values calculated for each point in the cycle. The baseline system assumes no temperature drop between the flow streams of the heat exchanger and an isentropic expansion in the expander that gives a 100% exergy efficiency for the refrigeration part of the system. In system A the temperature difference between the two fluid streams in the heat exchanger is taken as 5 °R; in system B, 10 °R; and in system C, 15 °R. This temperature difference is assumed to occur all along the heat exchanger with the cold inlet helium temperature remaining constant at 30 °R. A 60% isentropic efficiency is taken for the expander.

The results of the exergy analysis are given in Tables 5-15–5-18 for the different systems. The total heat-exchanger irreversible loss is divided between the two gas streams in these tabulations. Note the sharp increase in the heat-exchanger irreversible losses relative to the expander losses when the heat-exchanger temperature difference is increased. The irreversible heat-exchanger

Table 5-14 Operating Conditions of Reverse Brayton Cycle

Point	Pressure (psia)	Temperature (°R)	Enthalpy, h (Btu/lb)	Entropy, s (Btu/lb °R)	Exergy, Ex (Btu/lb)
			Reference Condition		
	14.7	530	664	7.480	0
			Baseline (Ideal, Reversible)		
1	161.7	540	678	6.318	630
2	145.5	30	40	2.696	1911
3	17.6	13	21	2.696	1892
4	17.6	30	43	3.820	1318
5	14.7	543	681	7.509	1.5
			System A		
1	161.7	540	678	6.318	630
2	145.5	35	47	2.912	1804
3	17.6	22	33	3.425	1518
4	17.6	30	43	3.820	1318
5	14.7	538	674	7.498	0.5
			System B		
1	161.7	540	678	6.318	630
2	145.5	40	54	3.094	1714
3	17.6	25	38	3.663	1397
4	17.6	30	'43	3.820	1318
5	14.7	537	667	7.485	0.3
			System C		
1	161.7	540	678	6.318	630
2	145.5	45	61	3.250	1639
3	17.6	29	42	3.770	1344
4	17.6	30	43	3.820	1318
5	14.7	527	660	7.472	0.3

Table 5-15 Baseline (Ideal, Reversible) System

	Units per lb/hr Helium Flow				Exergy Losses						Total Exergy Loss (%)
Station or Component	Exergy	Change in Exergy	Heat Transfer	Useful Work	Chemical Change	Heat Transfer	Friction	Machine Inefficiency	Heat Rejection	Total	
1	630										
Heat exchanger		1279	638								
2	1911										
Turbine		19		19							
3	1892										
Refrigeration load		556	22								
4	1318										
Heat exchanger		1318	638								
5	0										
Compressor		630		−630							
1	630									0	100%

Refrigeration load
100 W = 341.3 Btu/hr

341.3 Btu/hr = 15.5 lb/hr helium flow rate

Input exergy $= 630 - 19 = 611$ Btu/lb
Exergy efficiency $= 100\%$ (ideal, reversible)
Power required $= \dfrac{611 \text{ Btu/lb} \times 15.5 \text{ lb/hr}}{3.413 \text{ Btu/W hr}^{-1}} = 2775$ W

Table 5-16 System A

	Units per lb/hr Helium Flow			Exergy Losses							
Station or Component	Exergy	Change in Exergy	Heat Transfer	Useful Work	Chemical Change	Heat Transfer	Friction	Machine Inefficiency	Heat Rejection	Total	Total Exergy Loss (%)
1	630										
Heat exchanger 2	1804	1174	631			71				71	17
Turbine 3	1518	286		14				272		272	66
Refrigeration load 4	1318	200	10								
Heat exchanger 5	1318	631			71				71	17	
Compressor 1	0	630		−630							
	630						142		272	414	100

$$\frac{341\ \text{Btu/hr}}{10\ \text{Btu/lb}} = 34\ \text{lb/hr helium flow rate}$$

$$\text{Input exergy} = 630 - 14 = 616\ \text{Btu/lb}$$

$$\text{Exergy efficiency} = \frac{200}{616} = 0.32$$

$$\text{Power required} = \frac{616 \times 34}{3.413} = 6136\ \text{W}$$

137

Table 5-17 System B

Station or Component	Units per lb/hr Helium Flow					Exergy Losses					
	Exergy	Change in Exergy	Heat Transfer	Useful Work	Chemical Change	Heat Transfer	Friction	Machine Inefficiency	Heat Rejection	Total	Total Exergy Loss (%)
1	630										
Heat exchanger		1084	624			147				147	25
2	1714										
Turbine		317		16				301		301	50
3	1397										
Refrigeration load		79	5								
4	1318										
Heat exchanger		1318	624			147				147	25
5	0										
Compressor	630	630		−630							
1	630										
						294		301		595	100

$$\frac{341.3 \text{ Btu/hr}}{5 \text{ Btu/lb}} = 68.3 \text{ lb/hr Helium flow rate}$$

Input exergy $= 630 - 16 = 614$ Btu/lb

Exergy efficiency $= \dfrac{79}{614} = 0.13$

Power required $= \dfrac{614 \text{ Btu/lb} \times 68.3 \text{ lb/hr}}{3.413 \text{ Btu/hr W}} = 12{,}287 \text{ W}$

Table 5-18 System C

Station or Component	Units per lb/hr Helium Flow			Exergy Losses							Total Exergy Loss (%)
	Exergy	Change in Exergy	Heat Transfer	Useful Work	Chemical Change	Heat Transfer	Friction	Machine Inefficiency	Heat Rejection	Total	
1	630										
Heat exchanger		1009	617			154				154	26
2	1639										
Turbine		295		19				295		295	48
3	1344										
Refrigeration load		26	1								
4	1318										
Heat exchanger		1318	617			154				154	26
5	0										
Compressor		630		−630				295		295	
1	630					308		295		603	100

$$\frac{341.3 \ \text{Btu/hr}}{1 \ \text{Btu/lb}} = 341.3 \ \text{lb/hr helium flow rate}$$

$$\text{Input exergy} = 630 - 19 = 611 \ \text{Btu/lb}$$

$$\text{Exergy efficiency} = \frac{26}{611} = 0.04$$

$$\text{Power required} = \frac{611 \ \text{Btu/lb} \times 341.3 \ \text{Btu/hr}}{3.413 \ \text{Btu/hr W}} = 61{,}100 \ \text{W}$$

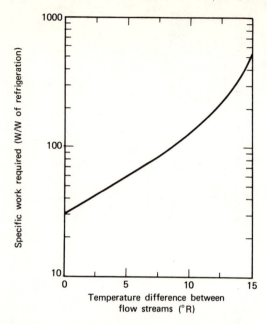

Figure 5-11.   Specific work required, reverse Brayton cycle.

loss becomes greater than the 60% efficient expander loss with only a 15°R temperature difference. This condition is equivalent to a heat-exchanger temperature effectiveness of 95%. The sharp increase in required power through the heat-exchanger irreversible loss is shown in Figure 5-11.

CHAPTER 6

Applications of Exergy Method

The exergy method of analysis has two advantages over the heat-balance method for design and performance analysis of energy-related systems. First, it provides a more accurate measurement of the actual inefficiencies in the system and the true location of these inefficiencies. It accomplishes this for any system, whether simple or complex. The exergy analysis also provides a true measure of the system efficiency for complex combined-cycle or open systems where the heat-balance method gives an erroneous efficiency value. In view of these factors, the areas of potential applications for the exergy method of analysis are reviewed in this chapter.

The application of the exergy analysis to a system should be guided by the need to improve the efficiency of energy use. These improvements must be acceptable economically as well as show gains in energy conservation. Present systems in operation should be analyzed with the exergy method to determine the areas where improvement in energy efficiency can be made effectively to conserve energy and reduce overal system operating costs. The design of future energy-related systems will emphasize efficiency and energy conservation to a greater extent than in the past. Systems will be more complex, tending to be an integration of combination of systems to maximize the use of the available energy resources. A major feature of these combined systems will be the increased transfer of heat between systems and the greater number of components that will involve irreversible losses. The efficiency of heat exchangers will become an even more critical factor than it is now in system design.

The design of energy-related systems will always be based on economics, designs will need to be more accurately evaluated, and consideration of life-cycle costs for fuel will be an important added factor in the analysis. The costs of irreversible losses of the available work in our energy systems can no longer be neglected in economic studies of industrial systems. This point has been recently emphasized by economists and others (Georgescu-Roegen 1975; Commoner 1977; Ross 1977).

6-1   ENERGY CONSERVATION

Conservation of our energy resources can be accomplished by increasing the efficiency of the way we use the available resources. To increase the efficiency of an existing process it is generally necessary to modify a piece of equipment or change the operating procedure that requires the outlay of capital expenditure. Therefore, it is important to determine the locations in the overall system where the maximum benefit in system efficiency is provided for the cost. To accomplish this, it is necessary to determine the performance characteristics of the working fluid at various stations in the system between processes and components. The losses involved in each process are then ascertained and their relationship to the overall system efficiency evaluated. From this knowledge, practical decisions for improving system efficiency can be developed.

The situation is somewhat different for the design of a new energy-related system. In this case we generally have the capability to parametrically analyze the performance and overall cost-effectiveness of different system configurations, different types of equipment, and methods of operation. It is important to correctly evaluate the losses in the system and their effect on the system efficiency when selecting the final design and operating conditions of a new system.

A major improvement in the efficient use of our energy resources can be achieved by directing efforts to the energy-intensive sections of our industry. This is where the greatest amount of the high-quality energy is used and where the most significant gains in energy conservation can be made. Because of the high temperatures and high pressures used in many industrial processes, these systems are primary candidates for recovery of intermediate and low-quality energy for use in other systems or for regeneration to improve the basic system efficiency. Improvements in this direction have been proposed by Berg (1974) and Sternlicht (1978).

To realistically perform these energy-recovery evaluations, the quality of the energy must be considered through use of exergy or available work analyses. It is important that the irreversible losses resulting from heat transfer be considered since energy recovery and regeneration usually involve heat exchangers.

Some potential-energy recovery techniques for heat and power systems involve the use of topping and bottoming cycles on the present convential cycles; replacing throttle valves with hydraulic turbines coupled to work devices; and waste-heat boilers and heat exchangers to provide steam, hot water, and hot air for required tasks such as supplementary processing or space heating. From an exergy-efficiency standpoint one of the most promising ways to use our high-quality energy is with a topping cycle to the conventional steam- or gas-turbine system. A magnetohydrodynamic cycle is an ideal method for accomplishing this, but the technology is still being developed toward its practical use. Closed-cycle liquid—metal systems such as with sodium or sodium—potassium alloy (NaK) have been proposed for topping cycle, but further development of this system is required for its acceptance for general use.

Another important aspect of energy conservation is to reduce losses of exergy or available work from the system. This can be done by replacing inefficient system components with modern, more efficient units. Heat and mass leakage from systems should be reduced as much as possible, especially for high-temperature and high-pressure regions where the energy quality is high.

The use of low-quality energy from alternate energy sources for heat and power has been given more serious consideration recently; and systems involving solar, wind, ocean thermal energy conversion, and geothermal energy are being designed, built, and operated. Since the available energy from these sources are low quality, the efficiency of the systems must be high to be economical. Irreversible losses cannot be neglected in the system design, or the overall system efficiency will be degraded below the economically required value. It should be pointed out that although systems that use low-quality alternate energy sources may be relatively costly for the work they perform, the economic evaluation of these systems should be made considering the cost of the conventional fuels they replace to perform the same task.

In the final analysis the basic principle for energy conservation while accomplishing the required work for our society is to match the quality of the energy with the quality required to accomplish the task. Low-quality tasks such as heating hot water or space heating and cooling should not be performed with high-quality energy resources, but rather with low-quality resources such as solar energy as far as possible. Where a high-quality task has been performed with a high-quality fuel such as in a fuel-fired gas-turbine—generator, the low-quality energy rejected from the system should be used to perform a low-quality task if at all practical.

The conservation measures discussed in the preceding paragraphs are generally well-known and accepted procedures. Many have been used in the past to various degrees. It is now necessary to implement these conservation measures extensively wherever there is need for improvements in our energy-use efficiency. For most effective application of energy-conservation efforts, the exergy-analysis approach should be used so that full consideration of system losses will be provided.

6-2   EXERGY ANALYSIS OF EXISTING SYSTEMS

Field tests performed on existing energy-related systems provide the data on which an energy balance is made using first-law principles, including the law of conservation of mass and energy. These data generally consist of temperatures, pressures, flow rates, voltages, velocities, and electrical currents at stations in the system between components and processes. When complete data are not available at a point in the system that would be desirable for a complete energy analysis, the lacking information can often be developed from other system data using the first-law energy balance. These energy-balance data are used as the basis for making an exergy balance of the system

The exergy balance provides the location and magnitude of the losses in work throughout the system that can be compared to losses that would occur for the system if modifications were made.

Modifications to an existing system can range from simple replacement of a component to the addition of a complete auxiliary system. Since the available work (exergy) in the initial energy supply is the real measure of total potential work that can be performed, the reduction in exergy loss by modification of the system can be directly related to the initial cost of energy. Cost savings in energy use are integrated with other life-cycle costs such as maintenance, capital cost, depreciation, and money-payback period in an overall economic evaluation of the impact of the modification. The decision to implement a modification can be based on a simple economic evaluation where the modification changes only the system losses and efficiency, but not the system configuration or end use.

Many energy-consumption programs for computer analysis are being developed to aid in energy-conservation studies of commerical and industrial systems. These programs usually are based on a first-law heat balance with overall empirically determined or assumed efficiencies for the components and processes. These computer programs can easily be upgraded to provide more realistic data on which to base decisions by adding exergy-analysis block subroutines to the program. This simple modification can account for irreversible losses in the system that are neglected or grossly assumed in most energy-balance analyses.

A modification to a base system that changes the system configuration or end use of the input energy results in a complex economic evaluation in which irreversible losses play a more dominant role. The reason for this is that changes to the basic system configuration usually involve transfer of energy from one circuit to another in which exergy losses are encountered. Bilder and Taubman (1973) consider the exergy loss in the contact heat exchanger used to couple a distillation system to the exhaust gas of a gas-turbine power system. Example 5-4 shows the effect on exergy loss and system exergy efficiency when the temperature difference between two circuits in a dual-cycle system is varied.

Systems of this type, combined cycles and cascaded cycles, will become more prevalent in the future as the economics of increased energy costs and decreased energy resource availability dictate.

6-3   EXERGY ANALYSIS OF NEW ENERGY-RELATED SYSTEMS

The application of exergy analysis to the design of new energy-related systems can be accomplished with the addition of the exergy-block calculations within or following the conventional energy-balance calculations. Figure 4-10 shows a typical flowchart of a system with the exergy-block calculations inserted within an energy-balance computer program.

A typical feature of a system-design program is to iterate on the operational conditions of specific components or to use the performance characteristics of different available commercial components to provide information that will support system-design decisions. The relative effect of the component and process inefficiencies on the cost of the output work of the system is necessary to make realistic decisions on the final system configuration and component selection. A comparison of exergy losses in the system with the cost of the input and output work will provide the true cost-effectiveness of individual system inefficiencies.

In addition to the system economic evaluation for minimum life-cycle cost, the evaluation of system design for energy conservation will be of interest in the future for planning purposes. This aspect of system design will become a controversial subject between free-enterprise business interests that favor minimum overall costs and maximum investor return and the energy conservationists and government regulators who will be promoting energy-resource conservation. It will be important in the evaluations made in this area that the quality of the energy involved be given consideration to assure a fair comparison. Expensive recovery of low-quality energy may not be in the best overall interest unless it is used to replace the existing expenditure of high-quality energy being inefficiently used for performing a task. It is tempting to assume that the large amount of low-quality energy generated by the sun, wind, and ocean can be harnessed for our future energy needs to replace oil, coal, and nuclear energy. The margin for error in the design of these low-quality energy systems is small, and even small, irreversible losses can significantly influence the actual performance of the system. Careful design considering the exergy losses in the system is required.

The design of new energy-related systems will involve the sharing of a high-quality energy source through cascading systems that will match the available quality of the energy with the required quality of the tasks. This combining of individual cycles and systems will require close examination of the irreversible losses in the energy-transfer components; otherwise, much of the effectiveness of cascading can be lost.

As the costs of our existing energy sources climb and desirable fuels become scarce in certain areas, the advantage of minimizing losses in the use of this energy will become more pronounced. The design of new energy-related systems to minimize the irreversible losses through an exergy analysis is a prime area for effective improvement of the efficient use of energy.

CHAPTER 7
Exergy Tables and Charts

This chapter contains tables and charts of exergy values for many of the common working fluids of energy-related systems. Additional tables and charts can readily be developed for any fluid for which enthalpy and entropy data are available by using the following equation for exergy:

$$Ex = (h - h_0) - T_0(s - s_0)$$

The reference values with the subscript 0 are selected for the fluid when it is at rest under ambient conditions that surround the system. A discussion on reference conditions for exergy analysis was given in Section 3-2.

For most exergy analyses of terrestrial energy systems, the reference "dead" state is the surrounding atmospheric environment. The tables and charts in this chapter are based on a reference condition of $70\,°F$ $(530\,°R)$ and $14.7\,psia$ where applicable. For many working fluids such as air and helium gas, this reference condition results in all values of exergy being positive for conditions other than the reference state as it should be. Exergy is a measure of available work and can not itself be negative.

Some fluids such as the Freon refrigerants do not have a well-defined tabulated state that matches the atmospheric conditions. On the other hand, these fluids are used in closed-cycle systems, and only the change in exergy of the refrigerant working fluid is of interest as it flows through the cycle. An approximate reference state is acceptable for most systems analyses using these types of fluid. A convenient reference for some fluids is saturated liquid at ambient temperature since there is negligible exergy involved in changing the pressure because there is little compressibility in liquids. The tables (Tables 7-1—7-10) and charts (Figures 7-1—7-10) in this chapter use $70\,°F$, $14.7\,psia$ vapor or $70\,°F$ saturated liquid for the reference state of these refrigerants, which results in some negative values of exergy. When used in a closed-cycle system the exergy values shown are used directly to provide the correct values of exergy change in a process or component in the system. The value of the fluid exergy relative to the reference state at 70 °F is the absolute value of the exergy at a given state regardless of the sign. This value is an approximation of the exergy value of the fluid relative to the surrounding environment.

Brodyanskii and Kalinin (Appendix b-3) discuss this problem of reference conditions and provide a method for correcting exergy values with change in the conditions of the surrounding environment.

The enthalpy and entropy values used in calculating the exergy values in these tables were taken from *Gas Tables* (Keenan 1945), the *Thermodynamic Properties of Refrigerants* (ASHRAE 1969). *Steam* (Babcock and Wilcox 1955), and NBS TN-622 (McCarty 1972).

Table 7-1   Exergy of Steam

Pressure (psia)	h (Btu/lb)	s (Btu/lb °R)	Ex (Btu/lb)
	Saturated Steam		
100	1186.6	1.6022	339.0
200	1197.8	1.5450	380.5
300	1202.4	1.5102	403.5
400	1204.1	1.4843	419.0
500	1203.7	1.4630	429.8
1000	1189.6	1.3881	455.4
1600	1162.7	1.3265	461.2
2000	1139.0	1.2896	457.0
2200	1123.8	1.2700	452.2
2600	1084.5	1.2257	436.4
3000	1025.6	1.1676	408.3
	Superheated steam 600 °F (1060 °R)		
100	1327.9	1.7569	398.3
200	1320.8	1.6756	434.3
300	1313.3	1.6256	453.3
400	1305.5	1.5884	465.2
500	1297.3	1.5579	473.2
600	1288.7	1.5316	478.5
800	1270.1	1.4862	484.0
1000	1248.7	1.4455	484.1
1200	1223.4	1.4058	479.9
1400	1192.4	1.3643	470.9

Table 7-1   Continued

Pressure	h (Btu/lb)	s (Btu/lb $^\circ$R)	Ex (Btu/lb)
	800 $^\circ$F (1240 $^\circ$R)		
100	1429.2	1.8445	453.2
200	1425.6	1.7662	491.0
300	1421.7	1.7193	512.0
400	1417.7	1.6854	526.0
500	1413.6	1.6586	536.1
1000	1391.0	1.5691	560.9
1500	1365.4	1.5091	567.1
2000	1337.0	1.4599	564.8
2400	1213.3	1.4244	588.9
3000	1271.1	1.3737	544.6
	1000 $^\circ$F (1640 $^\circ$R)		
100	1533.1	1.9209	516.6
200	1531.0	1.8438	555.3
300	1528.9	1.7983	577.3
400	1526.8	1.7658	592.5
500	1524.8	1.7404	603.9
1000	1514.8	1.6603	636.4
1500	1504.9	1.6120	652.1
2000	1494.7	1.5765	660.7
2400	1486.1	1.5531	664.5
3000	1472.9	1.5233	667.1

Reference conditions: $T_0 = 530 \,^\circ$R, $h_0 = 38.0$ Btu/lb, $s_0 = 0.746$ Btu/lb $^\circ$R. h and s values taken from *Steam, Its Generation and Use* with Permission from Babcock & Wilcox Company.

Table 7-2   Air at Low Pressure

T ($^\circ$R)	h (Btu/lb)	s (Btu/lb$^\circ$R)	Ex Btu/lb
100	23.74	0.1971	108.6
150	35.71	0.2958	68.3
200	47.67	0.3630	68.3
250	59.64	0.4164	228.3
300	71.61	0.4601	17.1
350	83.57	0.4970	9.53
400	95.53	0.5289	4.59
450	107.50	0.5571	1.62
500	119.48	0.5823	0.24
530	126.66	0.5963	0
550	131.46	0.6052	0.08
600	143.47	0.6261	1.02
800	191.81	0.6956	12.5
1000	240.98	0.7504	32.6
1500	369.17	0.8542	105.8
2000	504.71	0.9320	200.1
2500	645.78	0.9950	307.8
3000	790.68	1.0487	424.7
3500	938.40	1.0933	548.3
4000	1088.26	1.1333	677.1

Reference conditions: $T_0 = 530\,^\circ$R, $h_0 = 126.7$ Btu/lb, $s_0 = 0.5963$ Btu/lb$^\circ$R.

Table 7-3 Exergy of Anhydrous Ammonia

	Saturated liquid			Saturated Vapor		
T (°F)	h (Btu/lb)	s (Btu/lb °R)	Ex Btu/lb	h Btu/lb	s (Btu/lb °R)	Ex (Btu/lb)
−100	−63.3	−0.1626	165.5	572.5	1.6055	−135.8
−80	−42.2	−0.1057	156.4	581.2	1.5368	−90.7
−60	−21.2	−0.0517	148.8	589.6	1.4769	−50.6
−40	0.0	0.0	142.6	597.6	1.4242	−14.6
−20	21.4	0.0497	137.6	605.0	1.3774	17.6
0	42.9	0.0975	133.8	611.8	1.3352	46.7
20	64.7	0.1437	131.1	617.8	1.2969	73.0
40	86.8	0.1885	129.5	623.0	1.2618	96.8
60	109.2	0.2322	128.7	627.3	1.2294	118.3
80	132.0	0.2749	128.9	630.7	1.1991	137.8
100	155.2	0.3166	130.0	633.9	1.1705	156.1
120	179.0	0.3576	132.1	634.0·	1.1427	171.0

Superheated vapor

	75 psia				150 psia		
T (°F)	h (Btu/lb)	s (Btu/lb °R)	Ex (Btu/lb)	T (°F)	h (Btu/lb)	s (Btu/lb °R)	Ex (Btu/lb)
60	635.5	1.2839	97.6	100	645.9	1.2289	137.2
80	647.7	1.3071	97.6	140	672.3	1.2745	139.4
100	659.6	1.3286	98.0	180	696.9	1.3142	143.0
140	682.5	1.3682	99.9	220	720.7	1.3504	147.8
180	704.9	1.4044	103.2	260	744.3	1.3840	153.4
220	727.3	1.4380	107.8	300	767.7	1.4157	160.0
260	749.8	1.4705	113.0	340	791.2	1.4459	167.5
300	772.4	1.5011	119.4				

200 psia

T (°F)	h (Btu/lb)	s (Btu/lb °R)	Ex (Btu/lb)
100	635.6	1.1809	152.3
140	665.0	1.2317	154.8
180	691.3	1.2742	158.6
220	716.3	1.3120	163.5
260	740.5	1.3467	169.3
300	764.5	1.3791	176.2
340	788.5	1.4099	183.8
380	812.5	1.4392	192.3

Reference conditions: $T_0 = 530°R, h_0 = 653.7$ Btu/lb, $s_0 = 1.503$ Btu/lb °R.

Table 7-4 Exergy of Freon 12

T (°F)	h (Btu/lb)	s (Btu/lb °R)	Ex (Btu/lb)
Saturated liquid			
−140	−20.652	−0.056123	26.1
−120	−16.565	−0.043723	23.70
−100	−12.466	−0.032005	21.59
−80	−8.3451	−0.020862	19.82
−60	−4.1919	−0.010214	18.31
−40	0.0000	0.000000	17.09
−20	4.2357	0.009831	16.11
0	8.5207	0.019323	15.37
20	12.863	0.028515	14.84
40	17.273	0.037453	14.51
60	21.766	0.046180	14.38
80	26.365	0.054751	14.44
100	31.100	0.063227	14.68
140	41.162	0.080205	15.75
180	55.562	0.098039	20.69
200	59.203	0.10789	19.11
Saturated Vapor			
−140	61.896	0.2028	−28.50
−120	64.052	0.19359	−21.46
−100	66.248	0.18683	−15.66
−80	68.467	0.18143	−10.59
−60	70.693	0.17714	−6.08
−40	72.913	0.17373	−2.06
−20	75.110	0.17102	1.57
0	77.271	0.16888	4.90
20	79.385	0.16719	7.91
40	81.436	0.16586	10.61
60	83.409	0.16479	13.16
80	85.282	0.16392	15.51
100	87.029	0.16315	17.68
140	89.967	0.16159	21.47
180	91.561	0.15900	24.38
200	91.278	0.15651	25.43
Superheated Vapor			
3 psia			
−60	70.928	0.18709	−11.15
−40	73.463	0.19328	−11.90
−20	76.059	0.19932	−12.48
0	78.714	0.20523	−12.95
40	84.194	0.21665	−13.52

Table 7-4   Continued

T (°F)	h (Btu/lb)	s (Btu/lb °R)	Ex (Btu/lb)
80	89.889	0.22762	−13.64
120	95.785	0.23815	−13.32
160	101.869	0.24830	−12.62
200	108.127	0.25808	−11.58
		5 psia	
−60	70.729	0.17834	−6.68
−40	73.291	0.18459	−7.45
−20	75.909	0.19069	−8.07
0	78.582	0.19663	−8.52
40	84.090	0.20812	−9.11
80	89.806	0.21912	−9.22
120	95.717	0.22968	−8.93
160	101.812	0.23985	−8.19
200	108.079	0.24964	−7.11
		10 psia	
−20	75.526	0.17866	−2.09
0	78.246	0.18471	−2.55
40	83.828	0.19635	−3.12
80	89.596	0.20746	−3.28
120	95.546	0.21809	−2.95
140	98.586	0.22325	−2.61
180	104.793	0.23326	−1.76
200	107.957	0.23813	−1.14
		20 psia	
−20	−	−	−
0	77.550	0.17222	3.38
40	83.289	0.18419	2.76
80	89.169	0.19550	2.65
120	95.198	0.20628	2.95
140	98.270	0.21149	3.27
180	104.528	0.22159	4.16
200	107.712	0.22649	4.75
		100 psia	
100	88.694	0.16996	15.74
140	95.507	0.18172	16.30
180	102.257	0.19262	17.27
220	109.018	0.20287	18.58
260	115.828	0.21261	20.24
300	122.707	0.22191	22.20
340	129.665	0.23083	24.42

Table 7-4   Continued

T ($^\circ$F)	h (Btu/lb)	s (Btu/lb $^\circ$R)	Ex (Btu/lb)
		200 psia	
160	95.100	0.17130	21.40
200	102.652	0.18311	22.70
240	109.962	0.19387	24.29
280	117.178	0.20390	26.21
320	124.373	0.21337	28.36
360	131.583	0.22239	30.80
400	138.832	0.23102	33.49
		300 psia	
180	94.556	0.16537	23.99
200	98.975	0.17217	24.80
240	107.140	0.18419	26.61
280	114.879	0.19495	28.68
320	122.430	0.20489	30.93
360	129.900	0.21423	33.47
400	137.346	0.22310	36.20
440	144.800	0.23158	39.14
		400 psia	
200	93.718	0.16092	25.54
240	103.735	0.17568	27.71
280	112.286	0.18756	29.96
320	120.318	0.19814	32.42
360	128.112	0.20789	35.00
400	135.792	0.21704	37.87
440	143.424	0.22572	40.89
480	151.046	0.23401	44.12

Reference conditions: $T_0 = 70\,^\circ$F, $h_0 = 87.9$ Btu/lb, $s_0 = 0.1981$ Btu/lb $^\circ$R.
h and s values reprinted with permission from "Thermodrynamic Properties of Refrigerants," ASHRAE, 1969.

Table 7.5   Exergy of Freon 22

T (°F)	h (Btu/lb)	s (Btu/lb °R)	Ex (Btu/lb)
		Saturated liquid	
−140	−23.7	−0.0632	38.61
−120	−19.2	−0.0505	35.78
−100	−14.6	−0.0373	33.39
−80	−9.9	−0.0246	31.36
−60	−5.0	−0.0121	29.63
−40	0	0	28.22
−20	5.1	0.0119	27.01
0	10.4	0.0236	26.11
20	15.8	0.0350	25.47
40	21.4	0.0463	25.08
60	27.1	0.0574	24.90
80	33.1	0.0684	25.07
100	39.2	0.0794	25.34
120	45.7	0.0904	26.01
140	52.2	0.1016	26.57
160	59.9	0.1133	28.07
180	68.4	0.1263	29.68
		Saturated Vapor	
−140	86.7	0.2873	−37.35
−120	91.0	0.2739	−25.95
−100	93.4	0.2627	−17.61
−80	95.7	0.2534	−10.38
−60	98.0	0.2456	−3.95
−40	100.3	0.2389	1.90
−20	102.4	0.2332	7.02
0	104.5	0.2282	11.77
20	106.4	0.2238	16.01
40	108.1	0.2199	19.77
60	109.7	0.2163	23.28
80	111.0	0.2129	26.38
100	112.1	0.2096	29.23
120	112.8	0.2061	31.79
140	112.9	0.2024	33.85
160	112.3	0.1978	35.69
180	110.0	0.1913	36.83
		10 psia	
−40	100.690	0.24927	−3.21
−20	103.526	0.25588	−3.91
0	106.414	0.26230	−4.40
40	112.353	0.27468	−5.02

155

Table 7-5   Continued

T (°F)	h (Btu/lb)	s (Btu/lb °R)	Ex (Btu/lb)
80	118.512	0.28654	−5.12
100	121.674	0.29229	−5.03
150	129.822	0.30623	−4.25
200	138.310	0.31961	−2.87
250	147.129	0.33249	0.88
		20 psia	
0	102.785	0.23874	4.49
40	105.756	0.24535	3.97
80	111.826	0.25801	3.31
100	121.284	0.27588	3.27
150	129.510	0.28995	4.07
200	138.055	0.30342	5.47
250	146.916	0.31636	7.43
		100 psia	
60	110.700	0.22117	21.68
80	114.319	0.22801	21.68
100	117.911	0.23454	21.84
150	126.874	0.24988	22.64
200	135.931	0.26415	24.18
250	145.161	0.27764	26.25
300	154.605	0.29050	28.86
350	164.283	0.30283	32.02
		200 psia	
100	112.750	0.21165	28.82
150	123.133	0.22942	29.74
200	133.034	0.24503	31.40
250	142.826	0.25934	33.59
300	152.668	0.27274	36.36
350	162.638	0.28545	39.55
400	172.779	0.29760	43.27
		300 psia	
150	118.607	0.21441	33.19
200	129.784	0.23204	35.06
250	140.304	0.24741	37.40
300	150.621	0.26146	40.22
350	160.926	0.27460	43.58
400	171.314	0.28705	47.41
		400 psia	
160	115.516	0.20459	35.28
180	121.014	0.21332	36.17

156

Table 7-5   Continued

T (°F)	h (Btu/lb)	s (Btu/lb °R)	Ex (Btu/lb)
200	126.023	0.22104	37.09
250	137.549	0.23789	39.63
300	148.451	0.25273	42.69
350	159.144	0.26636	46.13
400	169.807	0.27914	50.10
450	180.534	0.29127	54.33

Reference conditions: $T_0 = 70\,°F$, $h_0 = 117\,Btu/lb$, $s_0 = 0.274$.
h and s values reprinted with permission from "Thermodynamic Properties of refrigerants," ASHRAE, 1969.

Table 7-6   Exergy of Freon 113

T (°F)	h (Btu/lb)	s (Btu/lb °R)	Ex (Btu/lb)	h (Btu/lb)	s (Btu/lb °R)	Ex (Btu/lb)
	Saturated Liquid			Saturated Vapor		
−30	1.97	0.0047	2.17	74.6	0.1738	−14.8
−20	3.96	0.0092	1.76	76.0	0.1732	−13.1
0	7.98	0.0182	1.01	78.9	0.1725	−9.9
20	12.0	0.0258	0.50	81.8	0.1722	−6.8
40	16.2	0.0352	0.18	84.6	0.1723	−4.0
60	20.4	0.0434	0.02	87.6	0.1728	−1.3
80	24.6	0.0515	0.01	90.5	0.1736	1.2
100	29.0	0.0594	0.19	93.4	0.1746	3.5
120	33.5	0.0673	0.49	96.4	0.1758	5.9
140	38.0	0.0750	0.98	99.4	0.1773	8.1
160	42.7	0.0827	1.58	102.3	0.1788	10.2
180	47.5	0.0903	2.35	105.2	0.1804	12.2
200	52.4	0.0978	3.29	108.1	0.1821	14.2
300	75.8	0.1297	9.75	120.2	0.1881	23.1
400	105.8	0.1659	20.5	122.6	0.1855	27.0
			Superheated Vapor			
		0.30 psia			0.50 psia	
−20	76.1	0.1770	−15.1			
0	78.9	0.1834	−15.6	78.9	0.1779	−12.7
20	81.8	0.1895	−16.0	81.8	0.1841	−13.1
40	84.7	0.1955	−16.2	84.7	0.1901	−13.4
60	87.7	0.2014	−16.3	87.7	0.1960	−13.5
80	90.8	0.2071	−16.3	90.8	0.2017	−13.5
100	93.8	0.2127	−16.2	93.8	0.2073	−13.4
150	101.7	0.2262	−15.5	101.7	0.2208	−12.6
200	109.7	0.2392	−14.4	109.9	0.2337	−11.3
250	118.4	0.2516	−12.2	118.4	0.2461	−9.3

Table 7-6   Continued

T (°F)	h (Btu/lb)	s (Btu/lb°R)	Ex (Btu/lb)	h (Btu/lb)	s (Btu/lb°R)	Ex (Btu/lb)
		1.0 psia			10.0 psia	
20	81.8	0.1768	−9.2			
40	84.7	0.1828	−9.5			
60	87.7	0.1886	−9.6			
80	90.7	0.1943	−9.6			
100	93.8	0.2000	−9.5	93.5	0.1751	3.4
150	101.7	0.2135	−8.8	101.4	0.1887	4.1
200	109.9	0.2264	−7.4	109.6	0.2017	5.5
250	118.4	0.2388	−5.5	118.1	0.2141	7.3
300	127.2	0.2507	−3.0	126.9	0.2261	9.8
350				136.0	0.2377	12.7
400				145.4	0.2489	16.2
		25 psia			50 psia	
150	100.9	0.1784	9.0			
200	109.2	0.1915	10.3	108.3	0.1833	13.8
250	117.7	0.2040	12.2	116.9	0.1959	15.8
300	126.5	0.2160	14.7	125.8	0.2081	18.2
350	135.7	0.2277	17.6	135.0	0.2198	21.2
400	145.0	0.2389	21.1	144.5	0.2311	24.7
450	154.7	0.2498	25.0	154.2	0.2421	28.6

Reference conditions: $t_0 = 70\,°F$, $h_0 = 22.5\ \text{Btu/lb}$, $s_0 = 0.0475\ \text{Btu/lb}°R$.

Table 7-7   Exergy of Freon 503

T (°F)	h (Btu/lb)	s (Btu/lb °R)	Ex (Btu/lb)
	Saturated Liquid		
−200	−51.7	0.4200	60.6
−180	−49.2	0.4291	58.2
−160	−46.5	0.4386	55.9
−140	−43.6	0.4481	53.8
−120	−40.2	0.4582	51.8
−100	−36.5	0.4689	49.8
−80	−32.4	0.4801	48.0
−60	−27.8	0.4918	46.4
−40	−22.6	0.5042	45.0
−20	−16.9	0.5173	43.8
0	−10.4	0.5313	42.9
20	−3.0	0.5465	42.2
60	18.6	0.5875	42.1

Table 7-7   Continued

T (°F)	h (Btu/lb)	s (Btu/lb °R)	Ex (Btu/lb)
		Saturated Vapor	
−200	24.7	0.7142	−19.0
−180	26.9	0.7013	−9.9
−160	29.1	0.6909	−2.2
−140	31.3	0.6822	−4.6
−120	33.4	0.6752	10.4
−100	35.6	0.6694	15.7
−80	37.6	0.6644	20.3
−60	39.4	0.6601	24.4
−40	41.1	0.6561	28.2
−20	42.4	0.6522	31.6
0	43.3	0.6482	34.6
20	43.6	0.6436	37.3
40	42.5	0.6373	39.6
60	35.9	0.6208	41.7
		10 psia	
−120	33.8	0.6898	3.1
−80	39.0	0.7045	0.5
−40	44.5	0.7183	−1.4
0	50.3	0.7314	−2.5
80	62.7	0.7561	−3.2
160	76.3	0.7798	−2.1
240	90.8	0.8016	0.8
320	106.0	0.8205	6.0
400	121.9	0.8345	14.5
480	137.9	0.8409	27.1
		20 psia	
−80	38.7	0.6882	8.8
−40	44.2	0.7021	6.9
0	50.1	0.7153	5.8
80	62.5	0.7401	5.1
160	76.1	0.7638	6.1
240	90.6	0.7857	9.0
320	106.0	0.8046	14.4
400	121.8	0.8187	22.7
480	137.8	0.8251	35.4
		100 psia	
−40	41.8	0.6617	25.9
−20	44.9	0.6689	25.2
0	48.0	0.6757	24.7
40	54.4	0.6889	24.1

Table 7-7 Continued

T (°F)	h (Btu/lb)	s (Btu/lb °R)	Ex (Btu/lb)
50	61.0	0.7015	24.0
120	67.8	0.7139	24.3
160	74.9	0.7258	25.1
200	82.2	0.7372	26.3
240	89.7	0.7480	28.1
280	97.3	0.7581	30.4
320	105.1	0.7672	33.3
360	113.1	0.7750	37.2
400	121.1	0.7814	41.8
500	141.3	0.7879	58.6
		150 psia	
−20	43.3	0.6571	29.9
0	46.6	0.6644	29.3
40	53.2	0.6780	28.7
80	60.0	0.6910	28.6
120	67.0	0.7036	28.9
160	74.1	0.7157	29.6
200	81.5	0.7272	30.9
240	89.0	0.7382	32.6
280	96.7	0.7483	34.9
320	104.6	0.7574	38.0
360	112.6	0.7654	41.8
400	120.6	0.7717	46.4
440	128.8	0.7762	52.3
500	140.9	0.7784	63.2
		200 psia	
0	45.0	0.6553	32.5
40	51.9	0.6696	31.9
80	58.9	0.6830	31.8
120	66.0	0.6959	32.0
160	73.3	0.7 82	32.8
200	80.8	0.7199	34.1
240	88.4	0.7309	35.9
280	96.2	0.7412	38.2
320	104.1	0.7412	46.1
360	112.1	0.7504	49.2
400	120.2	0.7648	49.7
440	128.4	0.7693	55.5
480	136.5	0.7714	62.5
500	140.6	0.7715	66.6

Table 7-7 Continued

T (°F)	h (Btu/lb)	s (Btu/lb °R)	Ex (Btu/lb)
	300 psia		
40	48.9	0.6560	36.1
80	56.5	0.6707	35.8
120	64.1	0.6842	36.3
160	71.6	0.6970	37.0
200	79.3	0.7090	38.4
240	87.1	0.7203	40.2
280	95.0	0.7308	42.5
320	103.0	0.7401	45.6
360	111.1	0.7482	49.4
400	119.3	0.7547	54.2
440	127.6	0.7593	60.0
480	135.8	0.7616	67.0
500	139.9	0.7617	71.0

Reference conditions: $T_0 = 60\,°F$, $h_0 = 60\,Btu/lb$, $s_0 = 0.745\,Btu/lb\,°R$.
h and s values reprinted with permission from "Thermodynamic Properties of Refrigerants," ASHRAE, 1969.

Table 7-8 Exergy of Methane

T (°R)	h (Btu/lb)	s (Btu/lb °R)	Ex (Btu/lb)	h (Btu/lb)	s (Btu/lb °R)	Ex (Btu/lb)
	Saturated liquid			Saturated vapor		
−280	−1934.0	1.087	492.3	−1705.8	2.357	47.4
−260	−1917.4	1.169	465.4	−1697.6	2.270	101.3
−240	−1900.5	1.244	442.6	−1690.7	2.198	146.8
−220	−1883.9	1.317	420.5	−1685.3	2.144	180.8
−200	−1865.8	1.387	401.5	−1681.0	2.098	209.5
−180	−1846.3	1.459	383.1	−1678.8	2.057	233.4
−160	−1826.2	1.527	367.1	−1679.2	2.018	253.7
−140	−1795.3	1.601	351.2	−1684.7	1.970	273.6
−120	−1757.6	1.730	327.9	−1698.0	1.905	294.8
	Superheated vapor					
	100 psia			500 psia		
−200	−1678.3	2.124	198.4			
−160	−1654.4	2.213	175.1			
−120	−1632.4	2.287	157.9	−1671.0	2.004	269.3
−80	−1611.0	2.350	145.9	−1637.2	2.101	251.6
−40	−1589.9	2.404	138.4	−1609.9	2.171	241.9
0	−1568.8	2.452	134.0	−1585.0	2.227	237.1

Table 7-8 Continued

T (°F)	h (Btu/lb)	s (Btu/lb°R)	Ex (Btu/lb)	h (Btu/lb)	s (Btu/lb°R)	Ex (Btu/lb)
40	−1547.6	2.496	131.9	−1561.1	2.277	234.5
80	−1526.1	2.536	132.2	−1537.6	2.321	234.7
120	−1504.1	2.574	134.2	−1514.1	2.362	236.4
160	−1481.6	2.612	136.4	−1490.4	2.401	239.9
200	−1458.6	1.648	140.3	−1466.3	2.439	243.7
400	−1332.9	2.810	180.1	−1337.5	2.609	282.6
600	−1188.1	2.965	243.0	−1190.8	2.761	348.1
800	−1023.9	3.108	331.2	−1025.3	2.905	437.8
1000				−843.0	3.040	548.0
		1000 psia			2000 psia	
−100	−1742.2	1.761	327.1	−1766.3	1.662	355.5
−60	−1661.9	1.978	292.1	−1720.8	1.785	335.4
−20	−1623.5	2.070	281.3	−1673.3	1.899	322.9
20	−1593.1	2.136	277.3	−1631.9	1.989	316.2
60	−1565.6	2.190	275.7	−1596.8	2.058	314.7
120	−1526.6	2.259	278.1	−1550.8	2.141	316.7
160	−1501.4	2.302	281.3	−1521.3	2.190	320.7
200	−1475.8	2.342	285.1	−1493.2	2.234	325.4
400	−1343.1	2.520	323.8	−1388.5	2.377	367.3
600	−1193.9	2.673	391.7	−1199.3	2.581	436.0
800	−1026.9	2.819	481.3	−1029.3	2.728	527.6
1000	−843.6	2.953	593.7	−844.3	2.861	642.1

Reference conditions: $T_0 = 530°R$, $h_0 = -1529.0$ Btu/lb, $s_0 = 2.780$ Btu/lb°R.

Table 7-9 Exergy of Isobutane

T (°F)	h (Btu/lb)	s (Btu/lb °R)	Ex (Btu/lb)
		Saturated liquids	
−20	−851.2	0.795	20.6
−10	−846.0	0.804	21.0
0	−840.7	0.8146	20.7
20	−829.8	0.8372	19.6
40	−818.4	0.8596	19.1
60	−806.8	0.8818	19.0
80	−795.1	0.9038	19.0
100	−782.9	0.9256	19.65
150	−751.0	0.9797	22.9
200	−715.3	1.0349	29.3
250	−673.5	1.0929	40.4

162

Table 7-9   Continued

T (°F)	h (Btu/lb)	s (Btu/lb °F)	Ex (Btu/lb)
		Saturated Vapor	
−20	−688.4	1.166	−13.3
−10	−685.1	1.162	−7.8
0	−681.8	1.1602	−3.6
20	−675.2	1.1590	3.6
40	−668.7	1.1590	10.1
60	−662.2	1.1595	16.4
80	−656.0	1.1611	21.7
100	−649.8	1.1632	26.8
150	−634.6	1.1705	29.2
200	−621.4	1.1773	47.8
250	−614.2	1.1765	55.4
		50 psia	
100	−647.0	1.180	20.7
140	−629.1	1.211	12.2
180	−610.4	1.241	25.0
200	−600.8	1.256	26.6
300	−549.3	1.328	40.0
400	−492.0	1.397	60.7
600	−362.9	1.534	117.2
		100 psia	
100	−653.8	1.147	31.4
140	−634.4	1.181	32.8
180	−614.8	1.212	36.0
200	−604.8	1.228	37.5
300	−552.1	1.302	51.0
400	−494.0	1.372	72.0
600	−364.1	1.510	128.7
		200 psia	
40	−818.2	0.8598	19.2
80	−794.7	0.9019	20.4
100	−782.7	0.9231	21.2
140	−757.7	0.9680	22.4
180	−625.7	1.176	44.1
200	−614.4	1.193	16.4
300	−558.1	1.272	60.9
400	−498.0	1.345	82.3
600	−366.6	1.484	140.0
		400 psia	
40	−818.1	0.8580	20.3
80	−794.5	0.9010	21.1

Table 7-9   Continued

T (°F)	h (Btu/lb)	s (Btu/lb °R)	Ex (Btu/lb)
100	−782.5	0.9226	21.6
140	−757.7	0.9660	23.4
180	−730.6	1.0101	27.2
200	−716.2	1.0351	28.3
300	−573.3	1.233	66.3
600	−371.8	1.457	149.1
800	−222.6	1.585	230.5

Reference conditions: $T_0 = 70\,°F$, $h_0 = -655.0\,Btu/lb$, $s_0 = 1.204\,Btu/lb\,°R$.
h and s values reprinted with permission from "Thermodynamic Properties of Refrigerants", ASHRAE, 1969.

Table 7-10   Exergy of Helium

T (°R)	h (Btu/lb)	s (Btu/lb °R)	Ex (Btu/lb)	h (Btu/lb)	s (Btu/lb °R)	Ex (Btu/lb)
		14.7 psia			30 psia	
100	130.5	5.414	561.2	130.5	5.059	749.3
200	254.7	6.274	229.6	254.8	5.920	417.3
300	378.8	6.778	86.6	378.9	6.423	274.8
400	502.9	7.135	21.4	503.1	6.781	209.3
500	627.0	7.412	0.2	627.2	7.058	186.6
600	751.2	7.638	3.2	751.3	7.284	190.9
800	999.0	7.995	61.8	1000.0	7.641	250.4
1000	1248.0	8.272	163.9	1248.0	7.916	352.6
1200	1496.0	8.498	292.2	1496.0	8.144	479.8
1400	1744.0	8.690	438.4	1744.0	8.338	625.0
1600	1992.0	8.855	599.0	1992.0	8.501	786.6
2000	2489.0	9.132	949.1	2489.0	8.778	1136.8
		100 psia			200 psia	
100	130.8	4.458	1068.2	131.1	4.110	1252.9
200	255.4	5.322	734.8	256.2	4.977	918.5
300	379.6	5.826	591.9	380.5	5.482	775.1
400	503.7	6.183	526.8	504.7	5.839	710.1
500	627.9	6.460	504.2	628.9	6.116	687.5
600	752.0	6.686	508.5	753.0	6.342	691.8
800	1000.0	7.043	567.3	1001.0	6.699	750.6
1000	1248.0	7.320	668.5	1249.0	6.976	851.8
1200	1497.0	7.547	797.2	1498.0	7.203	980.5
1400	1745.0	7.738	944.0	1746.0	7.394	1127.3
1600	1993.0	7.904	1104.0	1994.0	7.560	1287.3
2000	2490.0	8.181	1454.2	2490.0	7.837	1639.5

Table 7-10   Continued

T (°F)	h (Btu/lb)	s (Btu/lb °R)	Ex (Btu/lb)	h (Btu/lb)	s (Btu/lb °R)	Ex (Btu/lb)
		400 psia			600 psia	
100	131.8	3.758	1440.2	132.6	3.550	1551.2
200	257.7	4.632	1102.8	259.3	4.430	1211.5
300	382.3	5.138	959.3	384.2	4.936	1068.2
400	506.6	5.495	894.4	508.5	5.294	1002.8
500	630.8	5.772	871.7	632.7	5.571	980.2
600	754.9	5.999	875.5	756.9	5.798	984.1
800	1003.0	6.356	934.4	1005.0	6.155	1043.0
1000	1251.0	6.632	1036.1	1253.0	6.431	1144.7
1200	1500.0	6.859	1164.8	1501.0	6.658	1272.4
1400	1748.0	7.050	1311.6	1750.0	6.849	1420.1
1600	1996.0	7.216	1471.6	1998.0	7.015	1580.2
2000	2492.0	7.493	1820.8	2494.0	7.292	1929.3

Reference conditions: $T_0 = 530°R$, $h_0 = 664.3\,Btu/lb$, $s_0 = 7.480\,Btu/lb°R$.

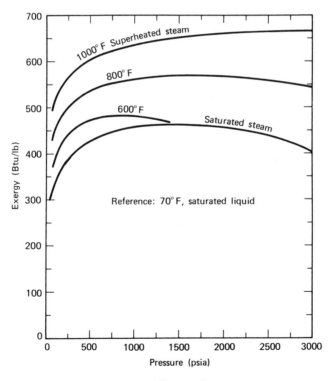

Figure 7-1.   Exergy of steam.

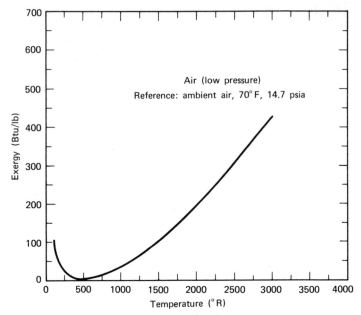

Figure 7-2. Exergy of air (low pressure).

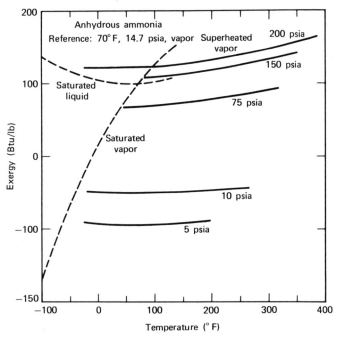

Figure 7-3. Exergy of anhydrous ammonia.

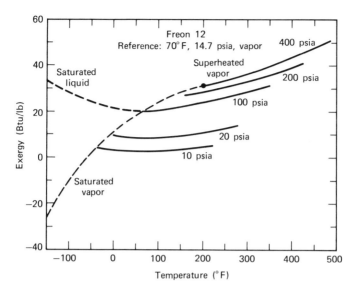

Figure 7-4. Exergy of Freon 12.

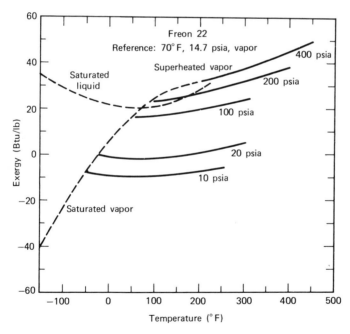

Figure 7-5. Exergy of Freon 22.

167

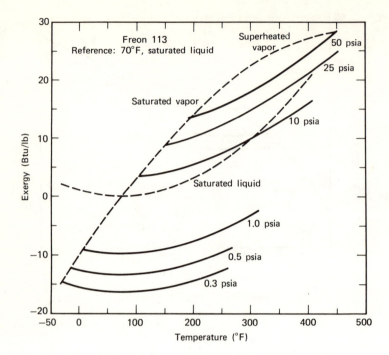

Figure 7-6. Exergy of Freon 113.

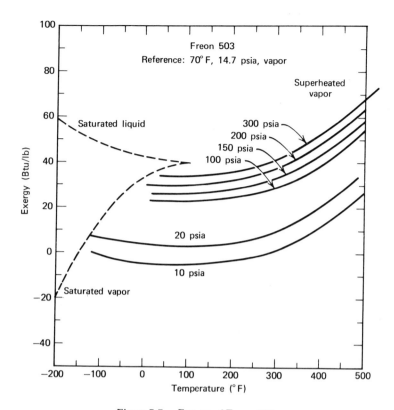

Figure 7-7. Exergy of Freon 503.

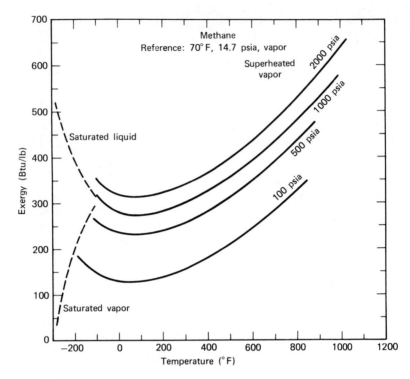

Figure 7-8. Exergy of methane.

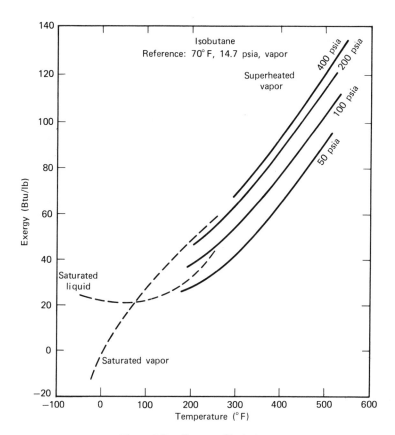

Figure 7-9. Exergy of isobutane.

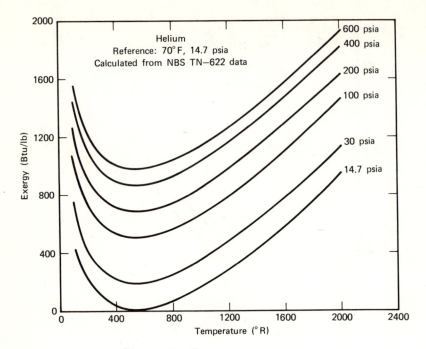

Figure 7-10. Exergy of helium.

Nomenclature and Conversion Constants

NOMENCLATURE

A	Area; availability, available work	PE	Potential energy
C	Thermal capacitance	Q	An amount of energy transferred as heat
c_p	Specific heat at constant pressure	q	Heat-transfer rate
c_v	Specific heat at constant volume	R	Thermal resistance; gas constant
E	Energy; emissive power	S, s	Entropy
e	Energy per unit of mass	T	Temperature
Ex, ex	Exergy	t	Time
F	Shape factor; force	U, u	Overall heat-transfer conductance; internal energy
f	Friction factor	u	Internal energy per unit mass
G	Flow stream mass velocity, W/A, V	V	Volume; velocity
		v	Volume per unit mass
g_c	Constant in Newton's law	W, w	An amount of energy transferred as work
g	Local acceleration caused by gravity	$\dot{w}$	Rate of work transfer
h	Enthalpy; heat-transfer coefficient	x	Quality of a two-phase mixture
k	Ratio of specific heats; thermal conductivity	z	Altitude coordinate
		γ	Specific heat ratio
L	Length	Δ	Denotes difference
M, m	Mass	ϵ	Emissivity; heat-exchanger effectiveness; temperature effectiveness
$\dot{m}$	Mass flow rate		
N, n	Number of species		
P, p	Pressure	η	Cycle efficiency

μ_j	Chemical potential of specie j	ρ	Density
μ	Viscosity	σ	Stefan–Boltzmann constant
		τ	Time

CONVERSION CONSTANTS

1 Btu/hr = 778.2 ft-lb
1 hp = 2545 Btu/hr
1 hp = 746 W
1 kW = 3413 Btu/hr
1 atm = 14.7 psia
1 atm = 760 torr
$^\circ C = (5/9)(^\circ F - 32)$

Bibliography (with References)

Adams, A. S., and Hilding, C. D., *Fundamentals of Thermodynamics,* Harper, New York, 1945.

Adkins, C. J., *Equilibrium Thermodynamics,* McGraw-Hill, London, 1968.

Allis, W. P., and Herlin, M. A., *Thermodynamics and Statistical Mechanics,* McGraw-Hill, New York, 1952.

American Society of Heating Refrigeration, and Air Conditioning Engineers (ASHRAE), *Thermodynamic Properties of Refrigerants,* 1969.

Andrews, F. C., *Thermodynamics: Principles and Applications,* Wiley-Interscience, New York, 1971.

Baehr, H. D., *Thermodynimik,* Springer-Verlag, Berlin, 1962 (in German).

Ball, W., "From Cascade Refrigerator for Liquid Air Plant Precooler," *Advances in Cryogenic Engineering,* Plenum, New York, 1960.

Baum, V. A., Khandurdyev, A., and Kakabaev, A., "A Study of a Solar Absorption Cooling Plant in Turkmenii During Summer," *Kholodi. Tekn.* (2) (1972) (in Russian).

Bejan, A., and Smith, J. L., Jr., "Thermodynamic Optimization of Mechanical Supports for Cryogenic Apparatus," *Cryogenics,* Vol. 14, pp 158–163 (Mar. 1974).

Bejan, A., "Discrete Cooling of Low Heat Leak Supports to 4.2 K," *Cryogenics* (May 1975) pp 290–293.

Bejan, A., and Smith, J. L., Jr., "Heat Exchangers for Vapor-Cooled Conducting Supports of Cryostats," *Advances in Cryogenic Engineering,* Vol. 20, Plenum, New York, 1975, pp. 247–256.

Bejan, A., "Two Thermodynamic Optima in the Design of Sensible Heat Units for Energy Storage," *ASME Journal of Heat Transfer,* 100, 708 (November 1978a).

Bejan, A., "General Criterion for Rating Heat Exchanger Performance," *Internat. J. Heat Mass Trans.,* 21, 655–658 (1978b).

Berry, C. W., *The Temperature-Entropy Diagram,* 3rd ed., Wiley, New York, 1911.

Berg, C. A., "A Technical Basis for Energy Conservation," *Mech. Eng.,* 30–42 (May 1974).

Bilder, Z. P., and Taubman, E. I., "Combining a Distillation Plant with a Gas Turbine," *Teploenergetika,* (2) (1973) (see Appendix B-10).

175

Bosnjakovic, F., *Technical Thermodynamics,* P. L. Blackshear, Jr., 3rd ed. Holt, Rinehart, & Winston, New York, 1965.

Bridgeman, P. W., "The Thermodynamics of Plastic Deformation and Generalized Entropy," *Rev. Mod. Phys.,* **22,** (1) (1950).

Callen, R. B., *Thermodynamics,* Wiley, New York, 1960.

Cravalho, E. G., McGrath, J. J., and Toscano, W. M., "Thermodynamic Analysis of the Regasification of LNG for the Desalination of Sea Water," *Cryogenics* (March 1977).

Commoner, B, *Business Week,* April 25, 1977, p. 69.

Daney, D. E., "Low Temperature Losses in Supercritical Helium Refrigerators," *Advances in Cryogenic Engineering,* Vol. 21, Plenum, New York, 1976, pp. 205–212.

Dean, J. W., and Mann, D. B., "The Joule–Thomson Process in Cryogenic Refrigeration Systems," National Bureau of Standards, NBS TN-227, February 1965.

DeBaufre, W. L., "Analysis of Power Plant Performance Based on the Second Law of Thermodynamics," *Mech. Eng.,* **47,** 426, 1925.

Donabedian, M., "Survey of Cryogenic Cooling Techniques," Aerospace report No. TR-00733901-01)-1, Air Force report No. SAMSO-TR-73-34, October 1972.

El-Sayed, Y. M., and Aplenc, A. J., "Application of the Thermoeconomic Approach to the Analysis and Optimization of a Vapor Compression Desalting System," *ASME, J. Heat Transf.* (May 1964).

Evans, R. B., "A proof That Essergy is the Only Consistent Measure of Potential Work (for Chemical Systems)," Ph.D. thesis, Dartmouth College, Hanover, N. H., June 1969.

Faires, V. M., *Applied Thermodynamics,* MacMillan, New York, 1938.

Gardner, J. B. and Smith, K. C., "Power Consumption and Thermodynamic Reversibility in Low Temperature Refrigeration and Separation Processes," *Advances in Cryogenic Engineering,* Plenum, New York, 1960.

Georgescu-Roeqen, N., *Business Week,* March 24, p. 109, 1975.

Gidaspow, D., "Hyperbolic Compressible Two-Phase Flow Equations Based on Stationary Principles and the Fick's Law," in *Two-Phase Transport and Reactor Safety,* Vol. 1, T. N. Veziroglu and S. Kakac, Eds., Hemisphere, Washington, D. C., 1978.

Green, M. A., Pines, H. S., and Doyle, P. A., "Computer Design and Optimization of Cryogenic Refrigeration Systems," *Cryogenics* (February 1979).

Hall, N. A., *Thermodynamics of Fluid Flow,* Prentice-Hall, Englewood Cliffs, N. J., 1951.

Hansen, R. E., "Irreversibility in the Theoretical Regenerative Steam Cycle," *Transact. ASME,* 557–560 (October 1945).

Haskin, W. L., and Dexter, P. F., "Ranges of Application for Cryogenic Radiators and Refrigerators on Space Satellites," AIAA paper No. 79-0179, 17th Aerospace Sciences Meeting, January 1979.

Howerton, M. T., *Engineering Thermodynamics,* Van Nostrand, Princeton, N. J., 1962.

Kalafati, D. D., *Thermodynamic Cycles of Nuclear Power Stations,* Gosenergizdat, Moscow, 1963 (English translation by Israel Program for Scientific Translation, Ltd.), 1965.

Kapitsa, P., "Expansion Turbine Producing Low Temperatures Applied to Air Liquefaction," in *Collected Papers of P. L. Kapitsa,* Vol II, Pergamon, Oxford, 1965 [originally presented in *J. Phys. USSR,* **1** (7), 1939].

Kays, W. M., and London A. L., *Compact Heat Exchangers,* McGraw-Hill, New York, 1958.

Keenan, J. H., *Thermodynamics,* Wiley, New York, 1941.

Keenan, J. H., "Availability and Irreversibility in Thermodynamics," *Br. J. Appl. Phys.* **2** (July 1951).

Keenan, J. H., and Kaye, J., *Gas Tables*, Wiley, New York, 1945.

Keller, A., "The Evaluation of Steam Power Plant Losses by Means of the Entropy Balance Diagram," *Transact. ASME* (October 1950), pp. 949–953.

Kun, L. C., and Ranov, T., "Efficiency of Low Temperature Expansion Machines," paper No. J-10, *Advances in Cryogenic Engineering*, Vol. 10, Plenum, New York, 1965.

Lansing, F. L., Strain, D. M., Chai, V. W., and Higgins, S., "The Updated Algorithm of the Energy Consumption Program (ECP) – A Computer Model Simulating Heating and Cooling Energy Loads in Buildings," DSN Progress report No. 42–49, November and December 1978, NASA/JPL, February 1979.

Lewis, G. N., and Randall, M., in *Thermodynamics*, revised by K. S. Pitzer and L. Brewer, Eds. McGraw-Hill, New York, 1961, chapter 7.

Littman, H., and Davidson, B., "Theoretical Bound of the Thermoelectric Figure of Merit from Irreversible Thermodynamics," *J. Appl. Phys.* **32** (2) 217 (1961).

Luikov, A. V., and Mikhailov, Yu. A., *Theory of Energy and Mass Transfer*, Pergamon, Oxford, 1965.

McCarty, R. D., "Thermophysical Properties of Helium-4 from 4 to 3000 R with Pressures to 15000 psia," National Bureau of Standards, NBS TN-622, September 1972.

McClaskey, B. M., and Lundquist, J. A., "Hydraulic Power Recovery Turbines," ASME paper No. 76-Pet-65, September 1976.

McKay, R. A., "Fuel Burner With Low Nitrogen Oxide Formation," NASA Technical Briefs, summer 1977, p. 257, NPO-13958.

Munster, A., *Classical Thermodynamics*, Wiley-Interscience, London, 1970.

Muhlenhaupt, R. C. and Strobridge, T. R., "An Analysis of the Brayton cycle as a Cryogenic Refrigerator," NBS TN 366, NBS, Cryogenic Division, August, 1968.

Obert, E. F., and Young, R. L., *Elements of Thermodynamics and Heat Transfer*, 2nd ed., McGraw-Hill, New York, 1962.

Oskima, K., Iishizaki, Y., Kamiyama, S., Akiyama, M., and Okuda, M., "The Utilization of LH$_2$ and LNG Cold for Generation of Electric Power by a Cryogenic Type Stirling Engine," *Cryogenics*, 617 (November 1978).

Perry, R. H., Chilton, C. H., and Kirkpatrick, S. D., Eds., *Chemical Engineers' Handbook*, 4th ed., McGraw-Hill, New York, 1963.

Petela, R., "Exergy of Heat Radiation," *ASME J. Heat Transf.*, 187–192 (May 1964).

Planck, M., *Theory of Heat*, Vol. V in *Introduction to Theoretical Physics*, H. L. Brose, Transl., Macmillan, London, 1932.

Progogine, I., *Introduction to Thermodynamics of Irreversible Processes*, 2nd ed., Interscience, New York, 1962.

Reynolds, W. C., and Perkins, H. C., *Engineering Thermodynamics*, McGraw-Hill, New York, 1970.

Ross, M., *Mech. Eng.*, 23 (July 1977).

Saha, M. N., and Srivastava, B. N., *A Treatise on Heat*, 4th ed., The Indian Press Private, Allahabad, 1958.

Schmidt, E., *Thermodynamics*, Oxford U. P., New York, 1949.

Selvey, A. M., and Knowlton, P. H., "Theoretical Regenerative Steam cycle Heat Rates," *Transact. ASME*, 489–503 (August 1944).

Smith, C. B., Ed., *Efficient Electrical Use*, Pergamon, New York, 1976 (see chapter 12 and Appendix C).

Sonntag, R. E. and Van Wynlan, G. J. *Introduction to Thermodynamics: Classical and Statistical*, Wiley, New York, 1971.

Steam, Babcock and Wilcox, New York, 1955.

Sternlicht, B., "Capturing Energy from Industrial Waste Heat," *Mech. Eng.*, 30–41 (August 1978).

Steward, F. R., and Guruz, H. K., "Mathematical Simulation of an Industrial Boiler by the Zone Method of Analysis," in *Heat Transfer in Flames*, N. H. Afgan and J. M. Beer, Eds. Halsted, New York, 1974.

Stoller, F. W., Lansing, F. L., Chai, V. W., and Higgins, J. S., "Energy Consumption Program – A computer Model Simulating Energy Loads in Buildings," NASA/Jet Propulsion Laboratory, DSN Progress report No. 42–45, March–April, 1978.

Tolman, R. C., *Relativity, Thermodynamics, and Cosmology*, Clarendon, Oxford, 1934.

Tolman, R. C., and Fine, P. C., "On Irreversible Production of Entropy," *Rev. Mod. Phys.*, **10** (1) (1948).

Trepp, C., "Refrigeration Systems for Temperatures Below 25°K with Turboexpanders," *Advances in Cryogenic Engineering*, Plenum, New York, 1961.

Tribus, M., *Thermostatics and Thermodynamics*, Van Nostrand, Princeton, N. J., 1961.

Tselikov, A. I., "Metal: Ways to Better Quality," *Nauka i Zhizn* (Science and Life) (1) (1976) (in Russian).

Van Wylen, G. J., *Thermodynamics*, Wiley, New York, 1959.

Van Wylen, G. J., and Sonntag, R. E., *Fundamentals of Classical Thermodynamics*, Wiley, New York, 1965.

Veinik, A. I., *Thermodynamics, A Generalized Approach*, Israel Program for Scientific Translations, Jerusalem, 1964.

Wepter, W. J., "First Word on Second Law," letter to the editor, *Mech. Eng.* (October 1977).

Worthing, A. G., and Halliday, D., *Heat*, Wiley, New York, 1948.

Zemansky, M. W., *Heat and Thermodynamics*, 4th ed., McGraw-Hill, New York, 1957.

Zimmerman, J. E., and Sullivan, D. B. "A Milliwatt Stirling Cryocooler for Temperatures Below 4 K," *Cryogenics* (March 1979).

Process Equation Blocks with Sample Calculations

Block A-1

Process: Combustion

Component: Furnace, combustor

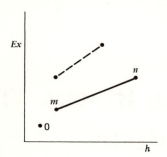 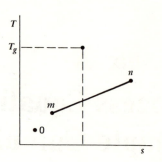

Reference conditions: T_0, s_0, h_0

	Point
$Ex_m = (h_m - h_0) - T_0(s_m - s_0)$	m

Exergy change – ideal performance

$$\Delta Ex' = (h'_n - h_m) - T_0(s'_n - s_m)$$

but $s'_n = s_m$

$$\Delta Ex' = (h'_n - h_m)$$

Exergy change – actual performance

$$\Delta Ex = (h_n - h_m) - T_0(s_n - s_m)$$

Exergy loss

$$Ex_{loss} = T_0(s_n - s_m) = T_0(s_n - s'_m)$$

	Point
$Ex_n = (h_n - h_0) - T_0(s_n - s_0)$	n

Block A-1a

Process: Steam generation, combustion fired

Component: Boiler, steam generator

Initial conditions, Water	Final Conditions, Superheated steam	Reference Conditions, Water
$102\,^\circ F$	$500\,psia,\ 800\,^\circ F$	$530\,^\circ R$
$h_1 = 70.0\ Btu/lb$	$h_2 = 1413.6\ Btu/lb$	$h_0 = 38.0\ Btu/lb$
$s_1 = 0.1326\ Btu/lb\,^\circ R$	$s_2 = 1.6586\ Btu/lb\,^\circ R$	$s_0 = 0.0754\ Btu/lb\,^\circ R$

$$Ex_1 = (70 - 38) - 530(0.1326 - 0.0754) = 1.68\ Btu/lb\ steam \qquad \text{Point 1}$$

Heat transferred into steam

$$1413.6 - 70 = 1343.6\ Btu/lb\ steam$$

Exergy loss

$$Ex_{loss} = 530(1.6586 - 0.1326) = 808.8\ Btu/lb$$

$$Ex_2 = (1413.6 - 38) - 530(1.6586 - 0.0754) = 536.5\ Btu/lb \qquad \text{Point 2}$$

Block A-2

Process: Combustion, chemical energy to heat

Component: Boiler, furnace, heater

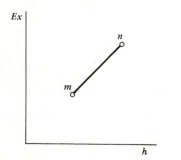 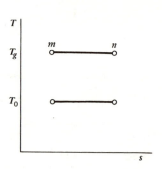

Reference conditions: T_0, s_0, h_0

	Point
$Ex_m = (h_m - h_0) - T_0(s_m - s_0)$	m

Exergy change – ideal performance

$$\Delta Ex' = 0$$

Exergy change – actual performance

$$\Delta Ex = (h_n - h_m)(T_0/T_g) = Q(T_0/T_g)$$

Exergy loss

$$Ex_{loss} = (h_n - h_m)(T_0/T_g) = Q(T_0/T_g)$$

	Point
$Ex_n = (h_n - h_0) - T_0(s_n - s_0)$	n

Block A-2a

Process: Combustion, chemical energy to heat energy

Component: Boiler, furnace, heater

Fuel conditions: 11.100 Btu/lb, 3960 °R adiabatic combustion temperature

Reference condition: 530 °R

	Point
$Ex_1 = 11{,}100$ Btu/lb fuel	1

Heat transferred from gas to steam

   11,100 Btu/lb fuel

Exergy loss – actual performance

   $Ex_{loss} = Q(T_0/T_g) = 11{,}100(530/3960) = 1490$ Btu/lb fuel

	Point
$Ex_2 = 11{,}100 - 1490 = 9610$ Btu/lb	2

Block A-3

Process: Combustion, heat transfer, gas to working fluid

Component: Boiler, furnace, heater

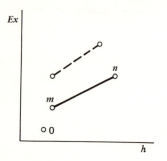 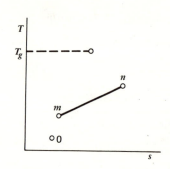

Reference conditions: T_0, s_0, h_0

	Point
$Ex_m = (h_m - h_0) - T_0(s_m - s_0)$	m

Exergy change — ideal performance

$\Delta Ex' = 0$

Exergy change — actual performance

$\Delta s = (s_n - s_m) - (h_n - h_m)/T_g$

$\Delta Ex = T_0 \Delta s$

Exergy loss

$Ex_{loss} = T_0[(s_n - s_m) - (h_n - h_m)/T_g]$

	Point
$Ex_n = (h_n - h_0) - T_0(s_n - s_0)$	n

Block A-3a

Process: Combustion, heat transfer, gas to working fluid

Component: Boiler, furnace, heater

Fuel: 11,100 Btu/lb

Adiabatic gas temperature: 3960 °R

Initial steam conditions: water, 84 °F, $h_1 = 52$ Btu/lb, $s_1 = 0.100$ Btu/lb °R.

Final steam conditions: superheated steam, 1000 psia, 700 °F, $h_2 = 1325$ Btu/lb, $s_2 = 1.5144$ Btu/lb °R.

Reference conditions: water, 70 °F, $h_0 = 38$ Btu/lb, $s_0 = 0.0746$ Btu/lb °R.

	Point
$Ex_1 = (52 - 38) - 530(0.100 - 0.0746) = 0.54$ Btu/lb steam	1

Exergy of gas stream: 12,000 Btu/lb fuel
Fuel: steam ratio: 8/1
Exergy transferred from gas: 1500 Btu/lb steam

$$Ex_{loss} = T_0 \left[(s_2 - s_1) - \frac{h_2 - h_1}{T_g} \right]$$

$$Ex_{loss} = 530 \left[(1.5144 - 0.100) - \frac{1325 - 52}{3960} \right]$$

$$= 579 \text{ Btu/lb steam}$$

	Point
$Ex_2 = (1325 - 38) - 530(1.5144 - 0.0746) = 524$ Btu/lb steam	2

Block A-4

Process: Frictional loss, vapor or gas flow

Component: Piping, ducts

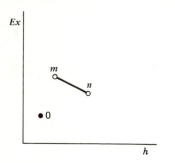 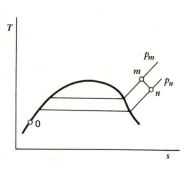

Reference conditions: T_0, s_0, h_0

	Point
$Ex_m = (h_m - h_0) - T_0(s_m - s_0)$	m

Exergy change – ideal performance

$\Delta Ex' = 0$

Exergy change – actual performance

$\Delta Ex = (h_m - h_n) - T_0(s_m - s_n)$

if $h_m = h_n$, $\Delta Ex = -T_0(s_m - s_n)$

Exergy loss

$Ex_{loss} = T_0(s_n - s_m) - (h_n - h_m)$

	Point
$Ex_n = (h_n - h_0) - T_0(s_n - s_0)$	n

Block A-4a

Process: Frictional loss, steam flow

Component: Steam piping

Initial Conditions	Final Conditions	Reference Conditions
800 psia, 1000 °F	700 psia, 1000 °F	70 °F, saturated water
$h_1 = 1511.0$ Btu/lb	$h_2 = 1513.9$ Btu/lb	$h_0 = 38.0$ Btu/lb
$s_1 = 1.6801$ Btu/lb °R	$s_2 = 1.6963$ Btu/lb °R	$s_0 = 0.0745$ Btu/lb °R

	Point
$Ex_1 = 1511 - 38 - 530(1.6801 - 0.745) = 622.0$ Btu/lb	1

Exergy change

$$\Delta Ex = (1513.9 - 1511) - 530(1.6963 - 1.6801) = 5.7 \text{ Btu/lb}$$

$$\Delta Ex = Ex_2 - Ex_1 = 622.0 - 616.3 = 5.7 \text{ Btu/lb}$$

Exergy loss

$$Ex_{loss} = 530(1.6963 - 1.6801) - (1513.9 - 1511) = 5.7 \text{ Btu/lb}$$

	Point
$Ex_2 = (1513.9 - 38) - 530(1.6963 - 0.0745) = 616.3$ Btu/lb	2

Block A-5

Process: Frictional loss, liquid flow

Component: Piping

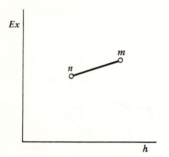 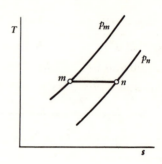

Reference conditions: T_0, s_0, h_0

	Point
$Ex_m = (h_m - h_0) - T_0(s_m - s_0)$	m

Exergy change — ideal performance

$\Delta Ex' = 0$

Exergy change — actual performance

$\Delta Ex = (h_n - h_m) - T_0(s_n - s_m)$

but $h_m = h_n$

$\Delta Ex = -T_0(s_n - s_m)$

Exergy loss

$Ex_{loss} = T_0(s_n - s_m) - (h_n - h_m)$

	Point
$Ex_n = (h_n - h_0) - T_0(s_n - s_0)$	n

Block A-5a

Process: Frictional loss, liquid flow

Component: Piping

Water

Condition 1	Condition 2
$P_1 = 2000\,\mathrm{psia}$	$P_2 = 400\,\mathrm{psia}$
$h_1 = 172.5\,\mathrm{Btu/lb}$	$h_2 = 165.9\,\mathrm{Btu/lb}$
$s_1 = 0.29143\,\mathrm{Btu/lb\,^\circ R}$	$s_2 = 0.29335\,\mathrm{Btu/lb\,^\circ R}$
$T_1 = 200\,^\circ\mathrm{F}$	$T_2 = 200\,^\circ\mathrm{F}$

Reference conditions: $T_0 = 530\,^\circ\mathrm{R}$, $h_0 = 38\,\mathrm{Btu/lb}$, $s_0 = 0.0745\,\mathrm{Btu/lb\,^\circ R}$

	Point
$Ex_1 = (172.5 - 38) - 530(0.29143 - 0.0745) = 19.5\,\mathrm{Btu/lb}$	1

Exergy change

$$\Delta Ex = (168.9 - 172.5) - 530(0.29143 - 0.29335) = 4.6\,\mathrm{Btu/lb}$$

$$\Delta Ex = Ex_1 - Ex_2 = 19.5 - 14.9 = 4.6\,\mathrm{Btu/lb}$$

Exergy loss

$$Ex_{\mathrm{loss}} = 530(0.29335 - 0.29143) - (168.9 - 172.5) = 4.6\,\mathrm{Btu/lb}$$

	Point
$Ex_2 = (168.9 - 38) - 530(0.29335 - 0.0745) = 14.9\,\mathrm{Btu/lb}$	2

Block A-6

Process: Heat loss, vapor or gas flow

Component: Piping

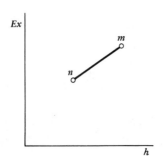
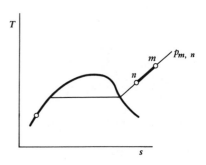

	Point
$Ex_m = (h_m - h_0) - T_0(s_m - s_0)$	m

Exergy change – ideal performance

$$\Delta Ex' = 0$$

Exergy change – actual performance

$$\Delta Ex = (h_m - h_n) - T_0(s_m - s_n)$$

Exergy loss

$$Ex_{loss} = T_0(s_n - s_m) - (h_n - h_m)$$

	Point
$Ex_n = (h_n - h_0) - T_0(s_n - s_0)$	n

Block A-6a

Process: Heat loss, vapor flow

Component: Steam piping

Initial conditions: superheated steam, $P_1 = 800$ psia, $T_1 = 1000\,°F$, $h_1 = 1511$ Btu/lb, $s_1 = 1.6801$ Btu/lb $°R$

Final conditions: superheated steam, $P_2 = 800$ psia, $T_2 = 900\,°F$, $h_2 = 1455.4$ Btu/lb, $s_2 = 1.6407$ Btu/lb $°R$

Reference conditions: saturated liquid, $T_0 = 70\,°F$, $h_0 = 38.0$ Btu/lb, $s_0 = 0.0745$ Btu/lb $°R$

	Point
$Ex_1 = (1511.0 - 38.0) - 530(1.6801 - 0.0745) = 622$ Btu/lb	1

Since no work is done in this process, all the exergy change is lost.

Exergy change

$$\Delta Ex = (1455.4 - 1511) - 530(1.6407 - 1.6801) = 34.7 \text{ Btu/lb}$$

Exergy loss

$$Ex_{loss} = 530(1.6407 - 1.6801) - (1455 - 1511) = 34.7 \text{ Btu/lb}$$

	Point
$Ex_2 = (1455 - 38) - 530(1.6407 - 0.0745) = 587.3$ Btu/lb	2

Block A-7

Process: Heat loss, liquid flow

Component: Piping

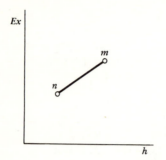 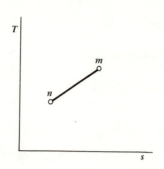

	Point
$Ex_m = (h_m - h_0) - T_0(s_m - s_0)$	m

Exergy change

$$\Delta Ex = (h_n - h_m) - T_0(s_n - s_m)$$

$$\Delta Ex = Ex_n - Ex_m$$

Exergy loss

If none of the heat loss is recovered, the exergy change in the process can be considered as the exergy loss

$$Ex_{loss} = T_0(h_n - h_m) - (s_n - s_m)$$

$$Ex_{loss} = Ex_n - Ex_m$$

	Point
$Ex_n = (h_n - h_0) - T_0(s_n - s_0)$	n

Block A-7a

Process: Heat loss, water flow

Component: Piping (Saturated Water)

Initial Condition	Final Condition
$T = 300\,^\circ\mathrm{F}$	$T = 200\,^\circ\mathrm{F}$
$h_1 = 269.6\ \mathrm{Btu/lb}$	$h_2 = 168\ \mathrm{Btu/lb}$
$s_1 = 0.4369\ \mathrm{Btu/lb\,^\circ R}$	$s_2 = 0.2938\ \mathrm{Btu/lb\,^\circ R}$

Reference conditions: $T_0 = 530\,^\circ\mathrm{R}$, $h_0 = 38\ \mathrm{Btu/lb}$, $s_0 = 0.0745\ \mathrm{Btu/lb\,^\circ R}$

	Point
$Ex_1 = (269.6 - 38) - 530(0.4369 - 0.0745) = 39.5\ \mathrm{Btu/lb}$	1

Exergy change

$$\Delta Ex = (269.6 - 168) - 530(0.4369 - 0.2938) = 39.5\ \mathrm{Btu/lb}$$

$$\Delta Ex = Ex_1 - Ex_2 = 25.8\ \mathrm{Btu/lb}$$

Exergy loss

Since none of the changes in exergy resulting from heat loss is recovered, the exergy loss is considered to be the change in the exergy

$$Ex_{\mathrm{loss}} = \Delta Ex = 39.5 - 13.7 = 25.8\ \mathrm{Btu/lb}$$

$$Ex_{\mathrm{loss}} = 530(0.2938 - 0.4369) - (269.6 - 168) = 25.8\ \mathrm{Btu/lb}$$

	Point
$Ex_2 = (168 - 38) - 530(0.2938 - 0.0745) = 13.7\ \mathrm{Btu/lb}$	2

Block A-8

Process: Flow heat and frictional loss, vapor or gas flow

Component: Piping, ducts

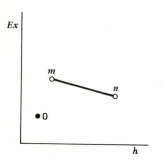
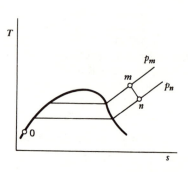

	Point
$Ex_m = (h_m - h_0) - T_0(s_m - s_0)$	m

Exergy change – actual performance

$$\Delta Ex = h_m - h_n - T_0(s_m - s_n)$$

$$\Delta Ex = Ex_m - Ex_n$$

Exergy loss

$$Ex_{loss} = T_0(s_n - s_m) - (h_n - h_m)$$

	Point
$Ex_n = (h_n - h_0) - T_0(s_n - s_0)$	n

Block A-8a

Process: Flow heat and frictional loss, steam flow

Component: Piping, ducts

Initial Conditions	Final Conditions	Reference Conditions
800 psia, 1000°F	700 psia, 900°F	530°F
Superheated steam	Superheated steam	Saturated liquid
$h_1 = 1511$ Btu/lb	$h_2 = 1459$ Btu/lb	$h_0 = 38$ Btu/lb
$s_1 = 1.680$ Btu/lb°R	$s_2 = 1.6573$ Btu/lb°R	$s_0 = 0.0745$ Btu/lb°R

	Point
$Ex_1 = (1511 - 38) - 530(1.680 - 0.0745) = 622$ Btu/lb	1

Exergy change

$$\Delta Ex = (1511 - 1459) - 530(1.680 - 1.6573) = 40 \text{ Btu/lb}$$

$$\Delta Ex = Ex_1 - Ex_2 = 622 - 582 = 40 \text{ Btu/lb}$$

Exergy loss

$$Ex_{loss} = 530(1.6573 - 1.68) - (1459 - 1511) = 40 \text{ Btu/lb}$$

	Point
$Ex_2 = (1459 - 38) - 530(1.6573 - 0.0745) = 582$ Btu/lb	2

Block A-9

Process: Flow heat and frictional loss, liquid flow

Component: Piping, ducts

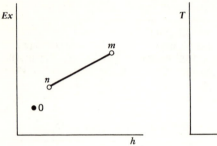 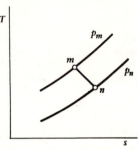

Reference conditions: T_0, s_0, h_0

	Point
$Ex_m = (h_m - h_0) - T_0(s_m - s_0)$	m

Exergy change – actual performance

$$\Delta Ex = (h_m - h_n) - T_0(s_m - s_n)$$

$$\Delta Ex = Ex_m - Ex_n$$

Exergy loss

$$Ex_{loss} = T_0(s_n - s_m) - (h_n - h_m)$$

	Point
$Ex_n = (h_n - h_0) - T_0(s_n - s_0)$	n

Block A-9a

Process: Flow heat and frictional loss, liquid flow

Component: Piping, ducts

Water Flow in a Pipe

Condition 1	Condition 2	Reference
$P = 1000\,\text{psia}$	$800\,\text{psia}$	
$T = 500\,^\circ\text{F}$	$300\,^\circ\text{F}$	$70\,^\circ\text{F}$
$h = 487.7\,\text{Btu/lb}$	$270.9\,\text{Btu/lb}$	$38.0\,\text{Btu/lb}$
$s = 0.6873\,\text{Btu/lb}\,^\circ\text{R}$	$0.43567\,\text{Btu/lb}\,^\circ\text{R}$	$0.0745\,\text{Btu/lb}\,^\circ\text{R}$

	Point
$Ex_1 = (487.7 - 38) - 530(0.68730 - 0.0745) = 124.9\,\text{Btu/lb}$	1

Exergy change

$$\Delta Ex = (487.7 - 270.9) - 530(0.68730 - 0.43567) = 83.4\,\text{Btu/lb}$$

$$\Delta Ex = Ex_1 - Ex_2 = 124.9 - 41.5 = 83.4\,\text{Btu/lb}$$

Exergy loss

$$Ex_{\text{loss}} = 530(0.43567 - 0.68730) - (270.9 - 487.7) = 83.4\,\text{Btu/lb}$$

	Point
$Ex_2 = (270.9 - 38.0) - 530(0.43567 - 0.0745) = 41.5\,\text{Btu/lb}$	2

Block A-10

Process: Pumping

Component: Liquid pump

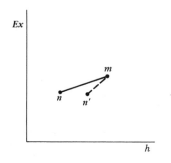 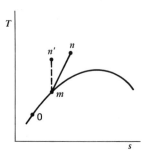

	Point
$Ex_m = (h_m - h_0) - T_0(s_m - s_0)$	m

Exergy change

The exergy change in this process involves the mechanical work expended in the flow process

$$\Delta Ex = \frac{v}{J}(P_n - P_m) + \frac{V_n^2 - V_m^2}{2gJ} + \frac{g}{g_c J}(z_n - z_m)$$

Flow   Momentum   Gravity
work

where $\frac{v}{J}(P_n - P_m) = (h_n - h_m) - T_0(s_n - s_m)$

Exergy loss

Since all the exergy change in this process is expended during the process, the exergy loss is the total exergy change

$$Ex_{loss} = Ex_n - Ex_m = \Delta Ex$$

	Point
$Ex_n = (h_n - h_0) - T_0(s_n - s_0)$	n

Block A-10a

Process: Pumping

Component: Liquid pump

This example involves pumping water from a zero elevation to an elevated tank that has a static pressure of 64.7 psia. The water is pumped from an open reservoir.

Initial Conditions	Final Conditions
$P = 14.7$ psia	$p = 64.7$ psia
$V = 0$ ft/sec	$V = 20$ ft/sec at tank inlet
Altitude: 0 ft	Altitude: 100 ft
Water flow rate: 75 gal/min (37,440 lb/hr)	

Point
1

Exergy change

$$\Delta Ex = \frac{1}{778}\left[0.016(144)(64.7 - 14.7) + \frac{20^2 - 0^2}{2(32.2)}\right.$$

$$\left. + \frac{32.2}{32.2}(100 - 0)\right] = 0.28 \text{ Btu/lb}$$

Total $\Delta Ex = (0.28)(37,440 \text{ lb/hr}) = 10,483$ Btu/hr

Exergy loss

Assuming no recovery of work during the pumping process,

$Ex_{loss} = Ex = 10,483$ Btu/hr $= 4.1$ hp

The 4.1 hp is required for this pumping process at 100% efficiency. If the pump is 80% efficient, the work required is $4.1/0.8 = 5.1$ hp.

Point
2

Block A-11

Process: Expansion with work, vapor

Component: Turbine

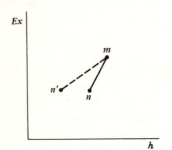
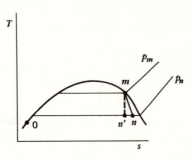

	Point
$Ex_m = (h_m - h_0) - T_0(s_m - s_0)$	m

Ideal performance

  Isentropic work $= h_m - h'_n$

Actual performance

  Work $= \eta\,(h_m - h'_n) = (h_m - h_n)$ actual work

Exergy change

  $\Delta Ex = (h_m - h_n) - T_0(s_m - s_n)$ total work

Exergy loss

  $Ex_{loss} = T_0(s_m - s_n)$ irreversible work (loss)

	Point
$Ex_n = (h_n - h_0) - T_0(s_n - s_0)$	n

Block A-11a

Process: Expansion with work

Component: Steam turbine

Initial Conditions	Final Conditions	Reference Conditions
$P = 500 \, \text{psia}$	$P = 3 \, \text{psia}$	$T_0 = 530\,^\circ\text{R}$
$T = 600\,^\circ\text{F}$		$h_0 = 38 \, \text{Btu/lb}$
$h = 1299 \, \text{Btu/lb}$		$s_0 = 0.0745$
$s = 1.5588$		
$x = 0.806$		

$$\eta = \text{Efficiency} = 75\%$$

$$Ex_1 = (1299 - 38) - 530(1.5588 - 0.0745) = 474 \, \text{Btu/lb}$$

Point 1

Ideal performance

$$h_2' = 109 + 0.806(1013) = 926 \, \text{Btu/lb}$$

$$\text{Work} = 1299 - 926 = 373 \, \text{Btu/lb isentropic work}$$

Actual performance

$$\text{Work} = 0.75(373) = 280 \, \text{Btu/lb actual work}$$

$$h_2 = 1299 - 280 = 1019 \, \text{Btu/lb}$$

$$x_2 = 1019 - 109/1013 = 0.898$$

$$s_2 = 0.2008 + 0.898(1.6855) = 1.7144$$

Exergy change

$$\Delta Ex = (1299 - 1019) - 530(1.5588 - 1.7144) = 361 \, \text{Btu/lb}$$

Exergy loss

$$Ex_{\text{loss}} = 530(1.7144 - 1.5588) = 81 \, \text{Btu/lb}$$

$$Ex_2 = (1019 - 38) - 530(1.7144 - 0.0745) = 113 \, \text{Btu/lb}$$

Point 2

Block A-12

Process: Expansion with work, gas

Component: Gas turbine

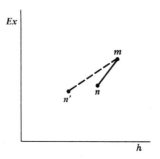 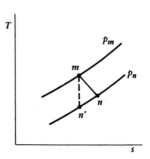

	Point
$Ex_m = (h_m - h_0) - T_0(s_m - s_0)$	m

Ideal performance

  assume $c_p = 0.24$ $T_n' = \dfrac{T_m}{(P_1/P_2)^{(k-1)/k}}$

  Isentropic work $= (h_m - h_n') = 0.24(T_m - T_n')$

Actual performance

  $T_n = T_m - \eta(T_m - T_n')$

  Actual work $= \eta(h_m = h_n') = (h_m - h_n) = 0.24(T_m - T_n)$

Exergy change

  $\Delta Ex = (h_m - h_n) - T_0(s_m - s_n)$ total work

Exergy loss

  $Ex_{loss} = T_0(s_m - s_n)$ irreversible work (lost)

	Point
$Ex_n = (h_n - h_0) - T_0(s_n - s_0)$	n

Block A-12a

Process: Expansion with work

Component: Gas turbine

Initial Conditions	Final Conditions	Reference Conditions
$P_1 = 88.2\,\text{psia}$	$P_2 = 14.7\,\text{psia}$	$p = 14.7\,\text{psia}$
$T_1 = 1500\,^\circ\text{F}, 1960\,^\circ\text{R}$		$T_0 = 530\,^\circ\text{R}$
$h_1 = 470\,\text{Btu/lb}$		$h_0 = 127\,\text{Btu/lb}$
$\theta = 0.92645$		$s_0 = 0.59630$
	$\eta = \text{Efficiency} = 70\%$	

$Ex_1 = (470 - 127) - 530(0.92645 - 0.59630) = 168\,\text{Btu/lb}$ Point 1

Ideal performance

$$T_2' = 1960/(88.2/14.7)^{0.286} = 1174\,^\circ\text{R}$$

Isentropic work $= 0.24(1960 - 1174) = 189\,\text{Btu/lb}$

Actual performance

$$T_2 = 1960 - 0.70(1960 - 1174) = 1410\,^\circ\text{R}$$

Actual work $= 0.24(1960 - 1410) = 132\,\text{Btu/lb}$

$$h_2 = 0.24(1410) = 338\,\text{Btu/lb}$$

Exergy change — actual

$$\Delta Ex = (470 - 338) - 530(0.92645 - 0.96072) = 150\,\text{Btu/lb}$$

Exergy loss

$$Ex_{\text{loss}} = 530(0.96072 - 0.92045) = 18\,\text{Btu/lb}$$

$Ex_2 = (338 - 127) - 530(0.96072 - 0.59630) = 18\,\text{Btu/lb}$ Point 2

Block A-13

Process: Condensation, saturated vapor to liquid

Component: Condenser

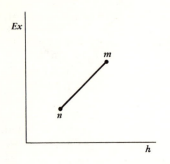 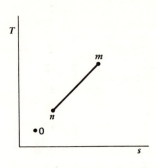

$$Ex_m = (h_m - h_0) - T_0(s_m - s_0)$$

Point
m

Exergy change

$$\Delta Ex = (h_n - h_m) - T_0(s_n - s_m)$$

$$\Delta Ex = Ex_m - Ex_n$$

Exergy loss

When the heat of condensation is not used to produce work, the total amount of exergy change is lost work; then

$$Ex_{\text{loss}} = Ex_m - Ex_n$$

$$Ex_{\text{loss}} = T_0(s_n - s_m) - (h_n - h_m)$$

$$Ex_n = (h_n - h_0) - T_0(s_n - s_0)$$

Point
n

Block A-13a

Process: Condensation, saturated vapor to liquid

Component: Condenser (steam condenser operating at 1.7 psia)

Initial Conditions	Final Conditions
$T = 300 \,^\circ F$	$T = 120 \,^\circ F$
$h = 1180 \, Btu/lb$	$h = 88 \, Btu/lb$
$s = 1.6350 \, Btu/lb \,^\circ R$	$s = 0.1645 \, Btu/lb \,^\circ R$

Reference conditions: $T_0 = 530 \,^\circ R$, $h_0 = 38 \, Btu/lb$, $s_0 = 0.0745 \, Btu/lb \,^\circ R$

Assume no recovery of heat of condensation

	Point
$Ex_1 = (1180 - 88) - 530(1.6350 - 0.0745) = 315 \, Btu/lb$	1

Exergy change

$$\Delta Ex = (1180 - 88) - 530(1.6350 - 0.1645) = 313 \, Btu/lb$$

$$\Delta Ex = Ex_1 - Ex_2 = 315 - 2 = 313 \, Btu/lb$$

Exergy loss

$$Ex_{loss} = 530(1.6350 - 0.1645) - (1180 - 88) = 313 \, Btu/lb$$

	Point
$Ex_2 = (88 - 38) - 530(0.1645 - 0.0745) = 2 \, Btu/lb$	2

Block A-14

Process: Heat transfer, different fluids

Component: Heat exchanger

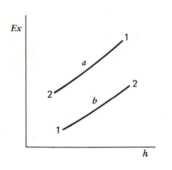

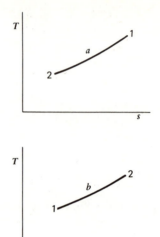

Different fluids cannot
be placed on same T–s
diagram.

$$Ex_{a_1} = (h_{a_1} - h_{a_0}) - T_{a_0}(s_{a_1} - s_{a_0})$$

$$Ex_{b_1} = (h_{b_1} - h_{b_0}) - T_{b_0}(s_{b_1} - s_{b_0})$$

Point
1

Heat balance – ideal performance

$$Q_a = m_a(h_{a_1} - h_{a_2})$$

$$Q_b = m_b(h_{b_1} - h_{b_2})$$

$$m_b = m_a(h_{a_1} - h_{a_2})/(h_{b_1} - h_{b_2})$$

Exergy change – actual performance

$$\Delta Ex_a = m_a[(h_{a_2} - h_{a_1}) - T_0(s_{a_2} - s_{a_1})]$$

$$\Delta Ex_b = m_b[(h_{b_2} - h_{b_1}) - T_0(s_{b_2} - s_{b_1})]$$

Exergy loss

$$Ex_{loss} = \Delta Ex_a - \Delta Ex_b$$

$$Ex_{a_2} = (h_{a_2} - h_{a_0}) - T_{a_0}(s_{a_2} - s_{a_0})$$

$$Ex_{b_2} = (h_{b_2} - h_{b_0}) - T_{b_0}(s_{b_2} - s_{b_0})$$

Point
2

Block A-14a

Process: Heat transfer, different fluids

Component: Heat exchanger

	Fluid a, Steam		Fluid b, Air	
	In	Out	In	Out
T	1000 °F	600 °F	70 °F	600 °F
P	500 psia	450 psia	14.7 psia	14.7 psia
h	1519.6	1302.8	127	256 Btu/lb
s	1.7363	1.5735	0.5960	0.7650 Btu/lb °R
Flow rate:	1 lb/hr		1.68 lb/hr	

Reference conditions: Steam, $T_0 = 530$ °R, $h_0 = 38$ Btu/lb,
$\qquad\qquad\qquad\qquad s_0 = 0.0745$ Btu/lb. °R
$\qquad\qquad$ air, $\quad T_0 = 530$ °R, $h_0 = 127$ Btu/lb,
$\qquad\qquad\qquad\qquad s_0 = 0.5960$ Btu/lb °R

$Ex_{a_1} = (1519.6 - 38) - 530(1.7363 - 0.0745) = 600.8$ Btu/lb

$Ex_{b_1} = (127 - 127) - 530(0.596 - 0.596) = 0$

Point 1

Air-to-steam weight flow = 1.68:1

Exergy change

Exergy given up by steam

$\quad \Delta Ex_a = (1519.6 - 1302.8) - 530(1.7363 - 1.5735) = 130.5$ Btu/lb steam

Exergy taken up by air

$\quad \Delta Ex_b = (256 - 127) - 530(0.7650 - 0.5960) = 39.4$ Btu/lb air

$\qquad$ or $(39.4)(1.68) = 66.2$ Btu/lb steam

Exergy loss

$\quad Ex_{loss} = \Delta Ex_a - \Delta Ex_b = 130.5 - 66.2 = 64.3$ Btu/lb steam

or

$\quad Ex_{loss} = \Sigma Ex_1 - \Sigma Ex_2 = 600.8 - 536.7 = 64.1$ Btu/lb steam

---

$Ex_{a_2} = (1302.8 - 38) - 530(1.5735 - 0.0745) = 470.5$ Btu/lb

$Ex_{b_2} = 1.68(256 - 127) - 530(0.7650 - 0.5960) = 66.2$ Btu/lb steam

Point 2

Block A-15

Process: Heat transfer, same fluid

Component: Heat exchanger

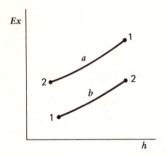
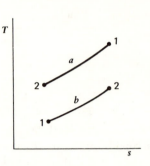

	Point
$Ex_m = (h_m - h_0) - T_0(s_m - s_0)$	m

Heat balance – ideal performance

$$\Delta h_a = h_{a_1} - h_{a_2}$$

$$\Delta h_b = h_{b_1} - h_{b_2}$$

Exergy change – actual performance

$$\Delta Ex_a = (h_{a_2} - h_{a_1}) - T_0(s_{a_2} - s_{a_1})$$

$$\Delta Ex_b = (h_{b_2} - h_{b_1}) - T_0(s_{b_2} - s_{b_1})$$

Exergy loss

$$Ex_{loss} = \Delta Ex_a - \Delta Ex_b$$

$$Ex_{loss} = T_0(\Delta s_a - \Delta s_b)$$

	Point
$Ex_n = (h_n - h_0) - T_0(s_n - s_0)$	n

Block A-15a

Process: Heat transfer, same fluid

Component: Heat exchanger

	Fluid a		Fluid b	
	1	2	1	2
T =	$800\,^{\circ}$R	$600\,^{\circ}$R	$400\,^{\circ}$R	$600\,^{\circ}$R
h =	306 Btu/lb	256	206	256
s =	0.80876 Btu/lb $^{\circ}$R	0.76496	0.71323	0.76496

Reference conditions: $T_0 = 530\,^{\circ}$R, $h_0 = 127$ Btu/lb, $s_0 = 0.5963$ Btu/lb $^{\circ}$R

Fluid: air at 1 atm pressure

$Ex_{a_1} = (306 - 127) - 530(0.80876 - 0.5963) = 67$ Btu/lb

$Ex_{b_1} = (206 - 127) - 530(0.71323 - 0.5963) = 17.5$ Btu/lb

Point 1

Heat balance – ideal performance

$\Delta h_a = 306 - 256 = 50$ Btu/lb

$\Delta h_b = 256 - 205 = 50$ Btu/lb

Exergy change – actual performance

$\Delta Ex_a = (256 - 306) - 530(0.76496 - 0.80876) = -27.5$ Btu/lb

$\Delta Ex_b = (256 - 206) - 530(0.76496 - 0.71323) = 22.1$ Btu/lb

Exergy loss

Exergy given up by fluid a: 27.5 Btu/lb

Exergy taken up by fluid b: 22.1

5.4 Btu/lb

Exergy loss: 5.4/27.5 = 20% of available work lost

$Ex_{a_2} = (256 - 127) - 530(0.76496 - 0.5963) = 39.5$ Btu/lb

$Ex_{b_2} = (256 - 127) - 530(0.76496 - 0.5963) = 39.5$ Btu/lb

Point 2

Block A-16

Process: Mixing of fluids

Component: Fluids with different temperatures

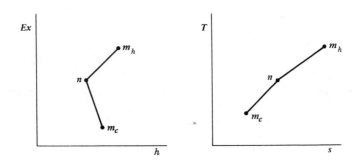

	Point
$Ex_m = (h_m - h_0) - T_0(s_m - s_0)$	m

Heat transferred

$$Q = w_c c_p (T - T_c) = w_h c_p (T_h - T)$$

$$T = \frac{w_c c_p T_c + w_h c_p T_h}{w_c + w_h}$$

$$\Delta s_i = w_i c_p \ln \frac{T_i}{T_i}$$

$$Ex_{loss} = T_0 \sum_i \Delta s_i$$

	Point
$Ex_n = (h_n - h_0) - T_0(s_n - s_0)$	n

Block A-16a

Process: Mixing of fluids

Component: Fluids with different temperatures

$w_c = 100 \, \text{lb/hr}$	$w_h = 200 \, \text{lb/hr}$	$T_0 = 530 \, ^\circ\text{R}$
$c_p = 0.24$	$c_p = 0.24$	$h_0 = 127 \, \text{Btu/lb}$
$T_c = 610 \, ^\circ\text{R}$	$T_h = 810 \, ^\circ\text{R}$	$s_0 = 0.5963$
$h_c = 146 \, \text{Btu/lb}$	$h_h = 194 \, \text{Btu/lb}$	
$s_c = 0.6300$	$s_h = 0.6986 \, \text{Btu/lb} \, ^\circ\text{R}$	

$Ex_{1h} = (200)(194 - 127) - 530(0.6986 - 0.5963) = 2620 \, \text{Btu/hr}$

$Ex_{1c} = (100)(146 - 127) - 530(0.6300 - 0.5963) = 150 \, \text{Btu/hr}$

Point 1

$Q = 100 \times 0.24(T - 610 \, ^\circ\text{R}) = 200 \times 0.24(T - 810 \, ^\circ\text{R})$

$$T = \frac{24(610) + 48(810)}{72} = 743 \, ^\circ\text{R}$$

$$\Delta s_c = 24 \ln \frac{743}{610} = 4.73$$

$$\Delta s_h = 48 \ln \frac{743}{810} = \underline{-4.14}$$

$$\text{net } \Delta s = +0.59 \, \text{Btu/hr} \, ^\circ\text{R}$$

Heat transferred

$Q = wc_p \, \Delta T = 3192 \, \text{Btu/hr}$

$Ex_{loss} 530(0.59) = 312 \, \text{Btu/hr} \quad Ex_{loss} = 10\%$

$Ex_2 = (300)(178 - 127) - 530(0.6776 - 0.5963) = 2460 \, \text{Btu/hr}$

Point 2

Block A-17

Process: Compression

Component: Compressor

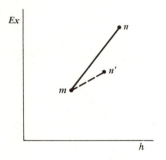 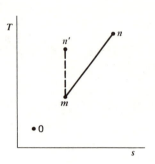

	Point
$Ex_m = (h_m - h_0) - T_0(s_m - s_0)$	m

Ideal performance

$\text{Required work} = h'_n - h_m = c_p(T'_n - T_m)$

$T'_n = T_m(P_n/P_m)^{(k-1)/k}$

Actual performance

$\text{Required work} = h_n - h_m = c_p(T_n - T_m)$

$T_n = T_m - (T'_n - T_m)/\eta$

Exergy change

Ideal performance

$\Delta Ex = h'_n - h_m$

Actual performance

$\Delta Ex = (h_n - h_m) - T_0(s_n - s_m)$

Exergy loss

$Ex_{\text{loss}} = T_0(s_n - s_m) = T_0(s'_n - s'_n)$

	Point
$Ex_n = (h_n - h_0) - T_0(s_n - s_0)$	n

Block A-17a

Process: Compression of air

Component: Compressor

Initial Conditions	Final Conditions	Reference Conditions
$P_1 = 14.7\,\text{psia},\ 70\,^\circ\text{F}$	$P_2 = 88\,\text{psia}$	$70\,^\circ\text{F},\ 14.7\,\text{psia}$
$h_1 = 127.2\,\text{Btu/lb}$		$h_0 = 127.2\,\text{Btu/lb}$
$s_1 = 0.7202\,\text{Btu/lb}\,^\circ\text{R}$		$s_0 = 0.07202\,\text{Btu/lb}\,^\circ\text{R}$

Compressor efficiency $\eta = 0.70$

Pressure ratio, 6 : 1

$k = 1.4$

$Ex_1 = 0$

Point 1

Ideal performance

$$T_2' = 530(88/14.7)^{0.286} = 530(1.67) = 885\,^\circ\text{R}$$

$$h_2 = 0.24(885) = 85.2\,\text{Btu/lb}$$

Work required $= 0.24(885 - 530) = 85.2\,\text{Btu/lb}$

Actual performance

$$T_2 = 530 + (885 - 530)/0.7 = 1037\,^\circ\text{R}$$

$$h_2 = 0.24(1037) = 248.9\,\text{Btu/lb}$$

Work required $= 0.24(1037 - 530) = 121.7\,\text{Btu/lb}$

Exergy change

Ideal

$$Ex = h_2' - h_1 = 212.4 - 127.2 = 85.2\,\text{Btu/lb}$$

Actual

$$Ex = h_2 - h_1 - 530(0.7595 - 0.7202) = 100.9\,\text{Btu/lb}$$

Exergy loss $= 530(0.7595 - 0.7202) = 20.8\,\text{Btu/lb}$

$Ex_2 = (248.9 - 127.2) - 530(0.7595 - 0.7202) = 100.9\,\text{Btu/lb}$

Point 2

Translations of Russian Papers on Exergy

Appendix B-1. Thermodynamic Analysis of Gas-Liquefaction Processes

Part I. Basic Methods of Analysis

V. M. BRODYANSKII

Inzhenerno-Fizicheskii Zhurnal, **6** (7) (July 1963)

A method of thermodynamic analysis for gas liquefaction is considered on the basis of exergy utilization. The classifications of the processes are presented and the basic method of separately determining the losses in the cooling processes during the liquefying part and during the heat-exchange part of the operation is described.

Liquefaction of gases as an independent process or as a component part of a more complex system for low-temperature separation of gas mixtures is widely used in technology. Until recently the region of applications for such processes extended to temperatures of 70–90°K, which is sufficient for liquefying air and its component parts and also methane, flourine, and carbon dioxide. Lower temperatures were used primarily in laboratory installations. Now the liquefaction of gases such as hydrogen, neon, and helium is developing into industrial scale, for which it is necessary to extend the region of working temperatures to 4–20°K.

The gas-liquefaction process requires the expenditure of significant energy, which rapidly increases with lowering of the temperature. This increase is determined not only by the increase in the theoretical minimal work by liquefaction, but by the rapid increase in the losses from irreversibility in different elements of the technical process. The thermodynamic efficiency of liquefaction plants ranges in the limits from 20% (for air and oxygen) to 5–10% (for helium), and the specific energy consumption reaches values that are characteristic of chemical plants with the same capacity. For example, to obtain one kilogram of magnesium by electrolysis, 16–22 kW-hr of energy is required; for phosphorous it is 14–20 kW-hr, for helium liquefaction, 25–30 kW-hr. It is necessary to raise the efficiency and to decrease the size and weight of the plants by development and to improve the methods of thermodynamic analysis of the liquefaction process.

There is a significant quantity of work in the literature devoted to thermal analysis [1–11]. In addition to the external energy characteristics, the losses through irreversibility have been studied in specific elements of the process. For this analysis the entropy method has been mostly used [1, 2, 7, 8, 10]; recently, the more satisfactory exergy method has been developed [9, 10].

However, the determination of the value for the losses through irreversibility is insufficient for a complete analysis of the process. A general method is necessary that permits finding the principles that determine the effectiveness of these processes and the factors that influence them. Basically, the exergy method can serve this purpose – the exergy method was developed for analyzing the specific processes in low-temperature technology [12–17] and cooling cycles [15–17].

IDEAL LIQUEFACTION PROCESSES

The process of converting a gas at ambient conditions ($T = 293\,°K$ and $p = 10^5\,N/m^2$) into a liquid at the boiling temperature and the original pressure is called *liquefaction*. In the ideal process of liquefaction all the expenditure of work goes to increasing the exergy (work capacity) of the gas [11, 12]; therefore, the minimum work required for liquefying 1 kg of gas is equal to the difference of the exergy (Ex) between the final and initial points independent of the path of the process.

From an infinitely large number of possible irreversible processes of liquefaction, three forms of the ideal process are selected that are effective for the technical methods of liquefaction under consideration.

The process of the first type, performed without a cooling cycle, gives a form in which the reversible output of energy to the surrounding environment is in the form of heat only at temperature T_0, the temperature of the environment. Consequently, the change in the state of the gas must be conducted only along the isotherm T_0 and the isotrope. Such a process was considered by Hausen [18] on a $T–s$ diagram by the cycle method. Figure B-1-1 shows this process in the coordinates $Ex–h$, which permits determination of the minimum work of liquefaction without additional diagrams [9, 11].

Beginning with the initial state (point 1), the compressed gas is processed isothermally to point 1, whose entropy is the same as the final point of the process (point 5). From pressure p_2 the gas is isentropically expanded to point 5 (line 2–5), which corresponds to complete liquefaction of the gas at pressure p_0, the pressure of the surrounding environment.

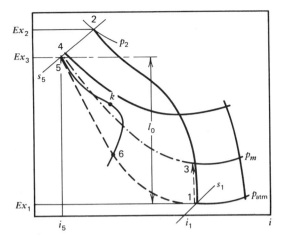

Figure B-1-1.   Depiction of ideal process of liquefaction by a diagram.

The expenditure of work on compression of the gas is equal to $Ex_2 - Ex_1$, part of which is recovered during expansion; the work of expansion is equal to $Ex_2 - Ex_5$ [12, 16]. The overall expenditure of work is determined by the equation

$$l_0 = (Ex_2 - Ex_1) - (Ex_2 - Ex_5) = Ex_5 - Ex_1 \qquad \text{(B-1-1)}$$

In technology, processes near to that described are used for liquefaction of gases that have critical temperatures near T_0 (e.g., chlorine, ammonia, and carbon dioxide gas). In this case the final pressure does not exceed the critical pressure p_{cr}.

For gases with a low critical temperature, for example, air, the pressure (p_2) must exceed the critical pressure. Therefore, the process of the first type is not used in technology for liquefaction of gases with low critical temperatures.

The process of the second type, performed without compression, is represented in Figure B-1-1 by line 1–6–5. The liquefying gas is isobarically cooled by supplementary cooling; in section 1–6 the cooling is done with a varying temperature and in section 6–5, with a constant temperature. The required work l_0, equal to $Ex_5 - Ex_1$, is expended completely by the cooling process.

In technology, liquefaction by methods near to the second type exists, for example, in the Philips gas-cooling machine.

The process of the third type occupies an intermediate position (figure B-1-1, lines 1–3–4–5). The gas being liquefied is subjected to pressurizing up to pressure p (point 3). The compressed gas is isobarically cooled to point 4 by means of an external cooling process, after which it is isentropically expanded to pressure p_0 (point 5).

The minimum work of liquefaction l_0 consists of two components.

For gas compression, the expended work is

$$l_0' = (Ex_3 - Ex_1) - (Ex_4 - Ex_5)$$ (B-1-2)

The work of the ideal cooling process is equal to its reduced cooling productivity q_0^{Ex} [15]:

$$l_0' = q_0^{Ex} = Ex_4 - Ex_3.$$ (B-1-3)

Figure B-1-2 is a graph showing values of l_0' for different gases as a function

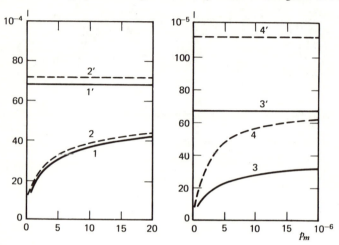

Figure B-1-2. Graphical functions of value of l_0' (1 and 2) and l_0 (1' and 2') on pressure p_m for air (1 and 1'), nitrogen (2 and 2'), helium (3 and 3'), and hydrogen (4 and 4').

of pressure. From the graph it is seen that for all gases when the pressure p_M is raised, the value of l_0' increases and consequently, l_0'' correspondingly decreases. For example, when $p = 10^6$ N/m² for air, l_0' is approximately 27% of l_0; when $p = 3 \times 10^6$, l_0' is around 40%; when $p = 20 \times 10^6$, l_0' reaches 63% of l_0. In the limiting cases when $p = p_2$, then $l_0 = l_0'$ (process of the first type) and when $p = p_0$, then $l_0 = l_0''$ (process of the second type).

INDUSTRIAL PROCESSES OF LIQUEFACTION

Most of the technical methods used for liquefaction are based on the third type — the simultaneous use of gas compression before liquefaction and supplementary cooling processes.

In the real case it is impossible to achieve a reversible isothermal compression 1–3 and isentropic expansion of the liquefying gas 4–5 (Figure B-1-1). Compression is done in the compressor where the process differs significantly from isothermal; the expansion takes place in a throttling valve. All these processes cause internal losses from the irreversibility d', which increases the required work up to a value of $l'_0 + d''$.

In the actual condition of removing heat from the liquefying gas to a working body in the cooling process there can be no reversibility; the presence of a finite-temperature difference leads to a supplementary loss, d'''. The lower is this loss, the better the cooling process matches the process of cooling the liquefying gas.

Therefore, the overall expenditure of work for the actual performance of the technical process is

$$l = l_0 + d' + l''_0 + d'' + d''' = (Ex_1 - Ex_5) + d' + d'' + d''' \qquad \text{(B-1-4)}$$

The proposed method of thermodynamic analysis is based on separately considering the losses d', d'', and d''' in the framework of the total process. This approach permits consideration of all gas-liquefying schemes operating with one or more cooling agents from a single point of view, and the approach can be used for studying the cooling-process methods that were developed for cooling-cycle analysis [14, 15].

We consider certain general principles connected with the thermodynamic losses d', d'', and d'''. The loss occurring through the insulation heat leak was not considered since its effect can be studied separately.

Losses in Liquefying Part of Gas

In all technical processes relating to the third type of liquefaction the gas goes through a change of state shown in the schematic and in the coordinates $Ex-i$ in Figure B-1-3. During the transfer of the gas into a liquid the gas always proceeds successively through specific processes.

Compression in Compressor. This occurs in conjunction with cooling to the discharge temperature (the condition shown by the isotherm 1–2). During compression the exergy of the gas increases to Ex_2 with the expenditure of work in the compressor. Under real conditions the process is irreversible, and part of the compressor work is lost by the amount d_K:

$$l_K = \Delta Ex_{2-1} + d_K \qquad \text{(B-1-5)}$$

Cooling in Heat Exchanger. This is accomplished by a cooling process (or processes). The exergy of the gas increases by ΔEx_{2-3} during the cooling process from 2 to 3. If the heat-exchanger resistance is neglected, the process of change in conditions can be considered to proceed without internal loss. (In certain cases the heat exchanger is divided into sections, and between these sections there are intermediate expansions of the gas in throttle valves or in expanders. In this case the corresponding losses must be considered in the analysis.)

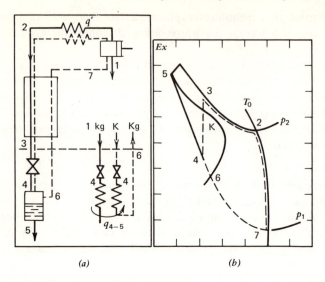

(a)   (b)

Figure B-1-3. Schematic of liquefaction process (*a*) and its depiction on the $Ex-i$ diagram (*b*).

Throttling. During this process (line 3—4) the pressure of the gas is reduced to p_1, and at the same time cooling occurs to T_4 with partial liquefaction. The losses from irreversibility during this process are

$$d_{\text{th}} = \Delta Ex_{3-4}.$$

Condensation. Condensation (process 4—5) proceeds because of the cooling process in which heat is removed from the liquefying gas $q_{4-5} = h_4 - h_5$. During this process the saturated vapor completely transforms into a liquid, and its exergy rises to Ex_5. It is necessary only to eliminate the limiting case when $h_3 = h_5$ since this is not achieved in practice as a rule. Therefore, a certain quantity of vapor is obtained after throttling (k kg/1 kg of liquid) that is separated from the liquid and used in the reverse stream for cooling the gas that is approaching the throttle valve. (The path of the vapor is shown in Figure B-1-3 as a dotted line.) The greater is the value of h_3, the greater is the quantity of this gas.

Therefore, the liquefaction process, which is independent of the presence of other cooling processes, always includes its own cooling cycle formed by the circulating part of the gas that passes through the process of compression, cooling and throttling along with the gas that is condensed into a liquid (Figure B-1-3*a*). In the Linde process there are no other cooling processes, and the rejection of heat in processes 2—3 and 3—4 is accomplished only with stream k. In the remaining cases part of the heat from the gas being reliquefied is removed

by a supplementary cooling system over some or all of section 2–3. The liquefaction in 4–5 is provided by the circulating part of the gas by transferring the
liquid from it to replace an equal quantity of dry saturated vapor in the gas
being liquefied.

This process, which proceeds in the separated liquid, is equivalent to reverse
heat exchange where 1 kg of saturated vapor is completely condensed (process
4–5 in Figure B-1-3), and k kg of saturated vapor is completely transformed
into a dry vapor (process 4–6) since $h_4 - h_5 = k(h_6 - h_4)$. All these processes
flow without loss. Such a representation in which the transfer of matter is
replaced by a reversible heat exchange (see the schematic in Figure B-1-3a)
leads to the same final results and permits considering the different changing
conditions in the liquefying gas and the cooling cycle that is a part of it.

Consequently, the loss d' in the liquefying part of the gas is made up of the
losses during compression in the compressor and the losses from throttling:

$$d' = d'_K + d'_{th} \qquad\qquad\qquad (B\text{-}1\text{-}6)$$

The value of d_K is related to the technical loss [13] and can be decreased
somewhat by improving the equipment. The value of d_{th} can be decreased only
by changing the parameters of the process (natural losses). The lower is the
temperature T_3, the higher is the efficiency of throttling η_{ex} and the lower is
d_{th} [12]. For example, for air the change in η_{ex} during expansion from
$(200-1) \times 10^5$ N/m^2 as a function of the initial temperature is characterized by
the following values:

Initial temperature ($^\circ$K)	172.5	150	130	120	110	100	90
Efficiency of throttling (η_{Ex}%)	36.6	53	67.4	72.8	78	81.1	83.4

By analogy, the analysis exists for other gases.

Losses in Cooling Processes

In the general case the cooling process is composed of two parts — the self-
cooling cycle with throttling and one or more auxiliary systems that can exist
in different forms. In the limit of liquefaction by the Linde method there are
no auxiliary systems. The auxiliary cooling processes usually involve either a
gas regenerator with an expander or a vapor cooling cycle with a different
coolant. The parameters of the basic and the auxiliary cycles are closely related.
Therefore, the losses in cooling process d'' are made up of two parts — the loss
in the main cycle with the throttling valve d''_m and the loss in the auxiliary
cycle d''_a. The factors determining these values must be analyzed in each specific
case by the methods developed for analyzing cooling cycles [15, 16].

Loss during Heat Exchange when Cooling Liquefying Gas

The cooling process must not only have a minimum value of d'', but must also correspond in the best way possible with the process of liquefaction so that the loss d'' during heat exchange with the liquefying gas in the process 2–3 will be minimal.

This loss is made up of two parts. The first (d'''_m) is connected to the heat transfer between the gas being liquefied in process 2–3 and the reverse stream in the main cooling process (the loss connected to the internal heat exchanger of the main cooling cycle in the regenerative heat exchanger is related to d''_m, and the second (d'''_a) is related to the thermal interaction of the gas being liquefied with the auxiliary cooling process.

The values of the losses d'''_m and d'''_a can be clearly represented on the $q–K_T$ diagram [15] shown in Figure B-1-4. The quantity of heat that is related to

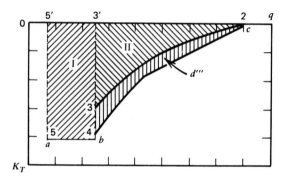

Figure B-1-4. Determination of heat-exchange loss by means of $q-K_T$ diagram.

1 kg of the gas being liquefied in processes 2–3 and 4–5 is plotted along the abscissa, and the values of K_T, where K_T is defined as $(T - T_0)/T$, are plotted along the ordinate. The line of heat transfer 2–3 is related to the cooling of the gas in the heat exchanger, and line 4–5 is related to the liquid-separation process. Area II is numerically equal to the reduced cooling production q^{Ex}_{2-3}, the minimal required for cooling. This value, equal to ΔEx_{2-3}, can also be directly determined from the $Ex–h$ diagram of the liquefying gas. By analogy, area I is equal to

$$q^{Ex}_{4-5} = \Delta Ex_{4-5}$$

Lines ab and bc are related to the cooling process. Line ab, coinciding with line 4–5 (the nominal dividing line), shows the change in K_T of the main cooling-cycle cooling agent in the separated liquid at constant temperature. There is no loss of exergy here. Line bc shows the variation of K_T of the coolant in the main and the supplementary process in the heat exchanger. Area $2-3'-b-c$ shows the reduced cooling production q_{bc}^{Ex} given up by the cooling process in the heat exchanger. If the temperature of the cooling agent is lower than that of the liquefying gas (and the absolute value of K_T is correspondingly greater), then

$$q_{bc}^{Ex} > q_{2-3}^{Ex}$$

The difference between these values corresponds to the area $2-3-b-c$ and is equal to the loss d'''. This value is distributed between the main and supplementary cooling process corresponding to the quantity of heat given up by them, and in total they equal q_{2-3}.

When the variation in the coolant temperature in the main auxiliary cycles differs, the lines of heat transfer are constructed separately for each section.

The value of q_{bc}^{Ex} can be calculated without constructing a $q-K_T$ diagram from the value of the exergy of the substances participating in the cooling process:

$$q_{bc}^{Ex} = kEx_m + Ex_{th} \qquad (7)$$

The first term on the right-hand side of the equation is related to the main cooling cycle and the second term, to the auxiliary cooling cycle.

As seen from Figure B-1-4, the change in the temperature of the cooling agent in the process of cooling the liquefying gas must be such that the value of

$$d''' = q_{bc}^{Ex} - q_{2-3'}^{Ex}$$

which expresses area $2-3-b-c$, will be minimal. This requires that a nonuniform condition be obtained with the smallest possible temperature difference along the heat exchanger since the same ΔT leads to different losses at different temperature levels.

SUMMARY

This appendix considers the method of thermodynamic analysis of gas liquefaction based on the exergy method. The processes are discussed, and the fundamentals of the methods by which the losses of cooling processes in a compressed portion of the gas may be defined both separately and with their thermal interaction are given.

REFERENCES

1. Keesom, W. H., *Commun. Leiden*, 76-a (1933).
2. Kaptisa, P. L., *ZhTF*, 9, 99 (1939).

 3. Hausen, H., *Zeitsch. Ges. Kalte-Ind.* (2)(1941).
 5. Hersh, S. Y., *Vestn. Mashino.* (11/12), 39 (1943).
 6. Kapitsa, P. L., *Vestn. Mashin.* (7/8), 43 (1944).
 7. Bliss, H., and Dodge, B., *Chem. Eng. Progr.*, 45, 33 (1949).
 8. Pavlov, K. F., and Malkov, M. P., *Cpravochnik po glubokomu okhlazhdeniyu* (Handbook of Cryogenics), Goskhmizdat, Moscow, 1947.
 9. Brodyanskii, V. M., and Meerzon, F. I., *Proisbodstvo kisloroda* (Production of Oxygen), Metallurgizdat, Moscow, 1947.
10. Malkov, M. P., Zeldovich, A. G., Gradkov, A. V., and Danilov, I. V., *Vydelenie deiteriya iz vodoroda metodom Glybokogo Okhlazhdeniya* (Separation of Deuterium from Hydrogen by Cryogenic Methods), Gosatomizdat, Moscow, 1961.
11. Baehr, H., *Chemie. Eng. Technol.*, 33, (5)(1961).
12. Ishkin, I. P., and Brodyanskii, V. M., *ZhTF*, 22, 1773 (1952).
13. Ishkin, I. P., and Brodyanskii, V. M., *ZhTF*, 22, 1783 (1952).
14. Ishkin, I. P., and Brodyanskii, V. M., *ZhTF*, 23, 1882 (1953).
15. Brodyanskii, V. M., and Ishkin, I. P., *Izvest. An SSR, OTN* (5), 1958.
16. Brodyanskii, V. M., and Medovar, L. E., *Kholod. Tekhn.* (5), (1961).
17. Brodyanskii, V. M., and Ishkin, I. P., *Kholodi. Tekhn.* (1)(1962).
18. Hausen, H., *Zeitsch. Ges. Kalte-Ind.* (7)(1925).

Appendix B-2. Thermodynamic Analysis of Gas-Liquefaction Processes

Part II. Analysis of Air Liquefaction by Linde Process

V. M. BRODYANSKII AND I. P. ISHKIN

Inzhenerno-Fizicheskii Zhurnal, **6** (10)(October 1963)

An analysis by the exergy method is presented for the process of air liquefaction by the Linde method. The values of the losses in different parts of the process are shown and the effects of the design factors determined.

The first useful process of gas liquefaction for industrial application was developed by K. Linde in 1895 and has remained important since one of the component parts [1] of all liquefaction processes is an initial gas compression.

Therefore, the Linde method of liquefaction has been selected for a basic exergy analysis using the method due to Brodyanskii et al. [1]. This analysis should clarify the principles involved in exergy analysis.

The schematic of the process is shown in Figure B-2-1. The liquefying part of the gas (1 kg) and the circulating part of the gas in the main cooling cycle (k kg) are shown separately. The liquefying part of the gas is separated from the circulating part into a discrete liquid. This process can be considered as a reversible heat exchange in which a quantity of heat $h_4 - h_5$ [1] is transferred from 1 kg of liquefying gas to the circulating part of the gas. Figure B-2-2 gives a depiction of the process of liquefaction in exergy–enthalpy coordinates separately for the liquefying gas (Figure B-2-2a) and the circulating gas (Figure B2-2b). Process $1-2-3-4-5$ is similar to that considered in Brodyanskii [1]; however, for all liquefying gas plants it is possible to separate out only the values of the loss d' that are composed of the losses during compression d_K and throttling d_{tv}.

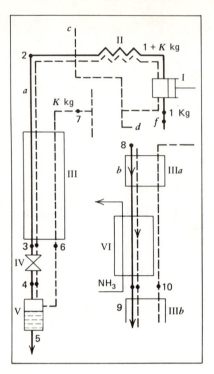

Figure B-2-1. Schematic of Linde liquefaction process: (*a*) without auxiliary cooling (I, Compressor; II, cooler; III, heat exchanger; IV, throttle valve; V, separated liquid); (*b*) with auxiliary cooling (IIIa, auxiliary heat exchanger; IIIb, main heat exchanger; IV, ammonia heat exchanger).

The main cooling cycle $1-2-3-4-6-1$ in the circulating part of the gas also enters into any process of liquefaction, deviating in each case only by the parameters. This process can be considered as a cycle since either k kg of gas enters the compressor from the atmosphere (the reservoir) or from the heat exchanger after being heated from T_7 to T_1 by the surrounding medium. The quantity of heat that is obtained by 1 kg of the circulating gas in process $7-1$ is equal to $h_1 - h_7$.

During compression of the gas in the cooling cycle its exergy increases by $\Delta Ex_{1-2} = k(Ex_2 - Ex_1)$. Further increase of the exergy in this part of the gas ΔEx_{2-3} occurs in the heat exchanger because of the decrease in the exergy of the reverse gas stream in process $6-7$. Throttling of the circulating part occurs the same as for the liquefying part of the gas. In process $4-6$ the exergy of the circulating gas decreases at the same rate as the exergy increases for the liquefying gas in process $4-5$, that is, by $k(Ex_4 - Ex_6) = Ex_5 - Ex_4$. During heating of the gas in process $7-1$ up to T_1 the exergy equal to $k(Ex_7 - Ex_1)$ is lost. This value is shown as a loss of exergy from the nonrecovery in the heat exchanger.

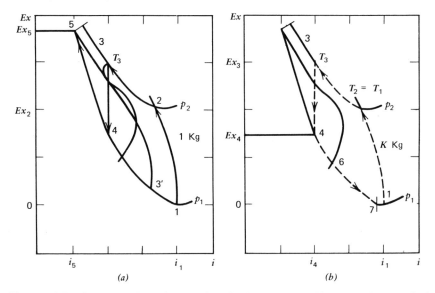

Figure B-2-2. Representation of Linde liquefaction process (without auxiliary cooling) in coordinates $Ex-i$: (a) liquefying part of gas (1 kg); (b) normal cooling cycle (k kg).

The exergy released in process 6–7 is used partially in the cycle for cooling the circulating part of the gas, and part of it is transferred to the liquefying gas cooled in process 2–3. The remainder is lost through irreversibility of the heat exchanger:

$$(Ex_6 - Ex_7)k = (Ex_3 - Ex_2)k + (Ex_3 - Ex_2) + d_T \qquad \text{(B-2-1)}$$

$$d_T = (\Delta Ex_{6-7} - \Delta Ex_{3-2})k - \Delta Ex_{3-2} \qquad \text{(B-2-1')}$$

The loss d_T is composed of two parts – the loss inside the cycle d_T'' and the loss during heat exchange between the liquefying gas and the cooling $d_T''' \therefore d_T = d_T'' + d_T'''$. Since the conditions for the heat-transfer cycle are the same in both cases,

$$\frac{d_T''}{d_T} = k \qquad \text{(B-2-2)}$$

Altogether, the work supplied in the compressor to the gas of the cooling cycle is expended on:

1. The exergy removed from the cycle for cooling the liquefying gas in the heat exchanger

$$(Ex_7 - Ex_6)\frac{k}{k+1}$$

This exergy, after deduction of the loss d_T''', is transferred to the liquefying gas.

2. The exergy removed in the separated liqued $(Ex_4 - Ex_6)k$, which is completely transferred to the liquefying gas.

3. The exergy loss resulting from the nonrecuperation $(Ex_7 - Ex_1)k$. The sum of the values of the exergy in 1, 2 and 3 make up the reduced cooling production of the cycle q_0^{Ex}.

4. The loss d'' from the irreversibility of the internal cycle composed of d_T'' (heat exchanger), d_{tv}'' (throttle valve), and d_K'' (compressor). Therefore, the algebraic sum of the change in exergy in the cycle is equal to zero:

$$\Delta Ex_{2-1} + \Delta Ex_{3-2} + \Delta Ex_{4-3} + \Delta Ex_{6-4} + \Delta Ex_{7-6} + \Delta Ex_{7-1} = 0$$

$$(3)$$

We performed a basic investigation relating to a quantitative analysis of the Linde process as an example of air liquefaction. The characteristics of the process obtained from the calculations are presented in Table B-2-1. Loss resulting from heat leakage through the insulation was not considered.

Table B-2-1   Characteristics of Air-Liquefaction Process by Linde Method ($\Delta T_{1-7} = 5^\circ$)

Characteristic	Point						
	1	2	3	4	; 5	6	7
Mass (kg)	12.6	12.6	12.6	12.6	1.0	11.6	11.6
Enthalpy (kJ/kg)	505.3	467.6	274.9	274.9	92.6	239.9	499.9
Pressure (10^{-5} N/m^2)	0.981	196.2	196.2	0.981	0.981	0.981	0.981
Temperature ($^\circ$K)	293.0	293.0	172.5	82.5	82.0	82.0	253.0
Exergy (kJ/kg)	0.0	427.8	488.5	199.4	685.1	157.5	0.126

On the basis of the data in Table B-2-1 and with the aid of the exergy—enthalpy diagram for air [4], all the basic properties of the process were calculated. The isothermal efficiency of the compressor η_E was taken as 0.65.

The overall expenditure of work in the compressor obtained from the calculations is $L_K = 8288.6$ kJ, and the work in the cooling cycle $L_0 = 7630.8$ kJ.

The basic values characterizing the process are given in Table B-2-2. The second column of Table B-2-2 shows the change in the exergy of the liquefying gas and the third column shows the exergetic heat given up by the cooling cycle (composed of the reduced cooling production in total). Columns 4, 5 and 6 correspond to the losses in the liquefying section d', the losses in the cooling cycle d'', and the losses during their thermal interaction d'''. Column 7 is the

Table B-2-2   Exergy Characteristics of Air-Liquefaction Process by Linde Method

Element or Process	$\Delta Ex'$	Ex_q''	d'	d''	d'''	Σd	η_{Ex}	θ
				Loss				
			kJ				%	
Compressor with cooler	427.8	–	230.0	2671.9	–	2901.9	65.0	50.1
Heat exchanger	58.7	144.5	0.0	1002.7	85.9	1088.8	40.8	12.0
Throttle valve	– 287.0	–	287.0	3329.4	–	3616.4	36.4	34.6
Liquid separation	485.6	485.6	0.0	0.0	0.0	0.0	100.0	3.2
Process 7–1	–	1.3	–	1.3	–	1.3	0.0	0.1
Total	685.1	631.4	517.0	7005.3	85.9	7608.2	–	100.0

total loss in each element of the process (Σd). The value of the exergetic efficiency of each element in the process is shown in column 8, and the exergetic importance of each element of the process θ, which is equal to the relationship of the excess exergy in each given element to the excess exergy in the whole cooling cycle, is shown in column 9 [4].

From comparison of the data obtained from the calculations, it is concluded that the low efficiency of the overal process (η_{Ex} = 8.3%) is caused by the small effectiveness of the main cooling cycle, which has around 92% of all the losses in Σd. The exergetic efficiency of the cycle

$$\eta_{Ex} = \frac{q_0^{Ex}}{L} = 8.3\%$$

is somewhat lower than conventional gas and vapor cooling cycles.

The loss d' in the liquefying part is comparatively small ($< 7\%$) even though the expansion is conducted with a high initial temperature and a relatively low efficiency (η_{Ex} = 36.4%). The loss during the thermal interaction of the gas in the cooling cycle and the liquefying gas d''' has practically no effect on the overall characteristics of the process ($\sim 1.1\%$) since the main part of the cooling production of the cycle (485.6 kJ) is delivered by the liquefying part of the gas to the separated liquid without loss.

We consider the distribution and the source of the losses in the cooling cycle. The most exergetic important item in the cycle is the compressor, (θ = 50.1%). Therefore, the loss in the compressor, in spite of the high efficiency η_{Ex} that it has compared to other elements in the cycle, reaches a significant value and shows a substantial effect of the expenditure of energy without changing the other parameters. As for every technical loss, it can be reduced to some extent by improving the equipment.

The second position for θ and the first for the absolute value of the loss is in the throttling valve (because of the low η_{Ex}). The heat exchanger is in third place in regard to the value of the loss. The loss in process 7–1 is so small that it has practically no significance.

The losses in the heat exchanger and during throttling are closely related, and for a given η_{Ex} of the compressor they determine the efficiency of the cooling cycle.

The function determining the loss in the heat exchanger is most satisfactorily found by using the $q-T$ and the $q-K_T$ diagrams [2, 3] presented in Figure

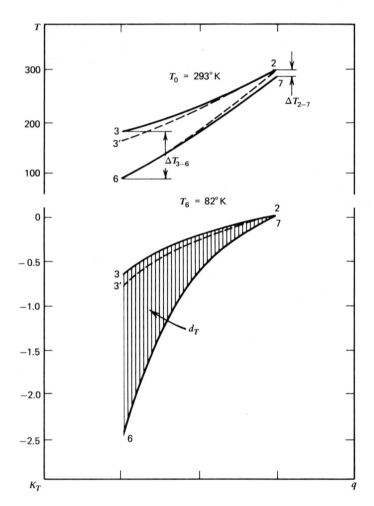

Figure B-2-3.   Heat exchanger $q-T$ and $q-K_T$ diagrams.

B-2-3. The upper part of the graph shows the temperature of the direct stream 2–3 and the reverse stream 6–7 in a given cross section of the heat exchanger as a function of the quantity of heat transferred by 1 kg of the liquefying air in the portion of the apparatus located between the cold end and the cross section under consideration. The temperature difference in any cross section of the heat exchanger is determined by the distance along the vertical between the two corresponding points on curves 2–3 and 6–7. In the lower part of the graph analogous curves are constructed, but along the ordinate axis the value of K_T is plotted instead of the temperature. According to the properties of the $q-K_T$ diagram [2, 3], area 2–3–6–7 in an appropriate scale is equal to the loss from irreversibility d relative to 1 kg of compressed air. The total loss in the heat exchanger d_T is equal to $d(k + 1)$, during which $d'' = dk$ and $d' = d$.

From the graph in Figure B-2-3 it is seen that the rise in temperature difference at the cold end of the heat exchanger leads to a sharp increase in the loss at the cold section. Part of the loss in the heat exchanger (technical loss) can be eliminated within limits if the surface area or its heat-transfer coefficient are increased so that $\Delta T_{2-7} \to 0$. As a result, the temperature difference at the cold end of the heat exchanger is also reduced. The flow of heat transfer in this case is shown by the dotted line in Figure B-2-3. The small change in the temperature difference at the cold end qf the heat exchanger (3.5 °) leads to a noticeable decrease in the value of d_T (5.6%). On the other hand, the decrease in the temperature difference at the warm end (5 °) has practically no effect on the loss in the apparatus. This difference is important only inasmuch as it determines the change in the temperature difference in that zone of the heat exchanger where thermal efficiency has significance. Lowering the temperature at point 3 leads to some increase in the efficiency of the throttling process and decreases d_{tv}. Calculations of the whole process for the conditions when $T_{7-2} \to 0$ show that the most significant change is in the value of k (from 11.6 to 9.9). This change decreases proportionally the total value of the losses in the throttling valve, heat exchanger, and compressor.

Therefore, the additional loss arising from the final temperature difference at the warm end of the heat exchanger is explained not by the loss of efficiency in cooling with the departing air, but by the increase in the loss d_T connected to both the rise in T_3 and the increase in the value of k. The loss through non-recuperation is only from the calculated efficiency and does not reflect the substance of the loss arising from the initial temperature difference.

The value of the specific terms in the external-energy balance of the cooling cycle can be clearly illustrated by comparing the resulting values expressed by energy and exergy.

Table B-2-3 shows a comparison of the exergy and the energy balances of the cooling cycle for the Linde process, including the external loss from heat exchange with the liquefying part of the gas. For clarity, only the low temperature

Table B-2-3   Energy and Exergy Balances for Processes in Linde Cycle

Energy Balance			Exergy Balance		
Form of Balance	kJ	%	Form of Balance	kJ	%
Cooling capacity			Exergy capacity		
$\Delta i_{1-2} k$	437.8	100	$\Delta Ex_{1-2} k$	4963.5	100
Cycle refrigeration used for liquefaction			Reduced refrigeration used for liquefaction		
$\Delta i_{2-3} + \Delta i_{4-5} = \Delta i_{2-6}$	375.0	85.4	$\Delta Ex_{3-2} + \Delta Ex_{5-4}$	544.3	11.0
Irreversible losses			Irreversible losses		
$\Delta i_{7-1} k$	62.8	14.6	$\Delta Ex_{7-1} k$	1.3	< 0.1
Other losses			Other losses		
Internal cycle	0.0	0.0	Internal cycle, d''	4332.0	87.2
External cycle for heat			External cycle for		
transfer	0.0	0.0	heat transfer,		
			d'''	85.9	1.8
Process efficiency	–	85.4	Process efficiency	–	11.0

part of the cycle is considered without compressors or coolers (line cd in Figure B-2-1 is the boundary of the system).

From comparison of the left and right parts of Table B-2-3 it is seen that the energy (thermal) balance in which the exergy of the heat is not considered shows the roles of the specific parts of the balance in a distorted form [2, 4]. The loss resulting from irreversibility generally is not reflected in the heat balance. Therefore, the effectiveness of the process with respect to heat balance is very high (85.4%), and the cooling is lost only through insufficient recuperation.

The actual percent of cooling lost through insufficient recuperation, as shown by the exergy balance, is near zero since the coefficient of work capacity at temperature T_7, which is near T_0, is less than 0.02. The effectiveness of the cycle is very low since a large part of the exergy is lost in the internal cycle (87%). Only 13% is transformed into cooling production, of which only 2% is lost during transfer between the circulating and liquefying parts of the gas.

The possibility of raising the effectiveness of the cooling cycle in the Linde process by means of reducing the technical losses can be accomplished only by increasing η_{Ex} of the compressor and by lowering ΔT_{7-1}. All that remains, the natural loss, cannot be decreased without substantially changing the process.

Any change of this type leads to, all things considered, decreasing the temperature at point 3 to a value more closely approaching T_6. Therefore, there would be a decrease in the losses in both the heat exchanger and in expansion valve of the main cooling cycle.

As an example, we consider how the efficiency of the Linde process changes when supplementary ammonia cooling is introduced at a temperature of $228\,°K$.

A achematic of the part of the process that changes with introduction of the auxiliary cooling is shown in Figure 3-2-1b (all the other resulting data — the compression pressure p_2, the compressor efficiency, and the temperature difference at the warm end of the heat exchanger — remain the same as in the previous case). The main cooling cycle undergoes significant change because part of its function is fulfilled by the ammonia cooling cycle. The temperature at point 3 in front of the throttling valve is lowered by 13.5% (to 159 °K), which leads to a rise in η_{Ex} of the throttling valve to 45.5%. The quantity of air circulating in the main cooling cycle is reduced to $k = 4.6 \, \text{kg/kg}$.

The stated changes in these factors of the loss resulting from the heat exchanger are seen from the diagram in Figure B-2-4. This diagram is different from Figure B-2-3 in that the plotted lines of heat transfer in Figure B-2-4 are all for the gas flow of k kg, rather than for 1 kg of air circulating in the cycle. The lines are constructed for two processes, with preliminary cooling and without it. In the first case $k = 4.6$, and in the second case $k = 11.6$. Consequently, the cross-hatched area on the diagram equals the corresponding value d_T'' for each case. From the graph it is seen how the loss in the heat exchanger decreases under the effect of decreasing ΔT and k. The area 8–9–10–10' equals the loss in the ammonia evaporator.

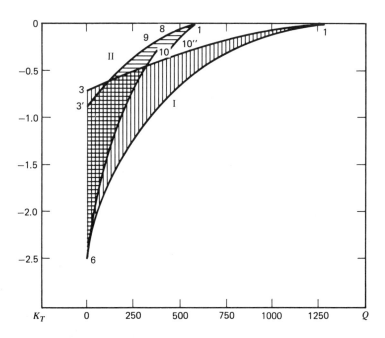

Figure B-2-4. $Q-K_T$ Diagram for process with (I) and without (II) auxiliary cooling.

The exergy efficiency of the main cooling cycle rises to 19.5% as a result of decreasing all the losses. The total expenditure of energy for liquefaction, including the ammonia equipment, correspondingly reduces to 213.9 kJ/kg, that is, more than a 200% reduction.

Further increase in the effectiveness of the main cooling cycle can be achieved by introducing the coolant at a lower temperature by means of either a vapor cooling cycle with another cooling agent or a gas cycle with one or more expanders.

The analysis is so important for use in technical liquefaction cycles that special consideration is merited for the described method.

The thermodynamic analysis permits establishing at what rate the auxiliary cooling is introduced (either vapor or gas) to achieve the goal of satisfying the requirements associated with the minimum expenditure of energy. The analysis must include not only the processes that are taking place in the main cooling cycle and in the liquefaction cycle, but also the complete characteristics of the auxiliary cycle.

The loss resulting from heat leakage from without can be sharply reduced in most cases by using newly perfected forms of insulation. In those cases when the effect of these losses must be considered in the analysis the required values can be easily determined by the described method if the quantity of heat and η_{Ex} of the actual cooling cycle are known. Often the use of calculating methods for the required supplementary work to reject heat flowing in the plant by means of the Carnot cycle leads to incorrect results.

SUMMARY

A distribution of losses and the role of some process elements are illustrated by the exergy analysis of air liquefaction by the Linde method. The role of changes appearing in the refrigeration cycle under the effect of additional cooling is shown.

REFERENCES

1.   Brodyanskii, V. M., *IFZh* (7)(1963).
2.   Brodyanskii, V. M., and Ishkin, I. P., *Izvest. AN SSSR, OTN* (5) (1958).
3.   Broadyanskii, V. M., and Medovar, L. E., *Kholodi. Techn.* (5)(1961).
4.   Ishkin, I. P., and Brodyanskii, V. M., *ZhTF*, (2)(1952).

Appendix B-3. Mass-Flow Exergy with Changing Parameters of Surrounding Medium

V. M. BRODYANSKII and N. V. KALININ

Inzhenerno-Fizicheskii Zhurnal, **10** (5)(May 1966)

Two methods are considered for determining the exergy value of the stream mass when the parameters of the surrounding medium are changed. The first method involves using an exergy diagram in dimensionless values, and the second method involves constructing a correction scale for the conventional diagram.

In many domestic and foreign reports on thermodynamic analysis using the exergy method the question repeatedly arises regarding the necessity to consider changes in the surrounding medium. Graphical and analytical methods exist to determine the correction in calculating the exergy of the stream mass during temperature change of the surrounding medium [1–10].

However, in certain regions of technology there is a significant change not only in the temperature, but in the pressure of the surrounding medium (flight at different altitudes and the operation of plants deep under the earth). It is sufficient to say that for a change in flight altitude from $H = 0$ km up to $H = 12$ km the temperature of the surrounding medium changes from $T_{H=0} = 293\,^\circ$K up to $T_{H=12} = 217\,^\circ$K, and the pressure changes from $P_{H=0} = 1$ atm up to $P_{H=12} = 0.21$ atm.

The change in the parameters of the surrounding medium is especially significant in evaluation of the operating effectiveness of similar plants and their elements in the different regimes [11, 12].

Examination of the exergy diagram shows that the effect of the parameters T_0 and P_0 on the exergy value of the stream is different. It has been shown [1, 2, 4, 5, 8] that for a deviation of the temperature T_0 from a standard value the sign and value of the correction is determined by both the value of the deviation and the parameters of the state of the mass for which the exergy value is calculated. The correction is introduced with the aid of specially constructed auxiliary lines.

In contrast, when p_0 is changed the value and the sign of the correction are identically determined by the deviation of the pressure of the surrounding medium from a standard value. During this the exergy value deviates between certain additive constants.

If an entropy diagram is lacking for the state of the working body, a diagram in dimensionless values is constructed from the equation of an ideal gas independent of the parameters of the surrounding medium [13] for approximately determining the exergy of the stream mass.

From the overall differential equation of exergy

$$dEx \; = \; dh - T_0 \, ds \tag{B-3-1}$$

we obtain the following equation by substituting the value of ds from the equation of an ideal gas:

$$dEx \; = \; dh - T_0 \left(c_p \, \frac{dT}{T} - R \, \frac{dp}{p} \right). \tag{B-3-2}$$

Integrating eq. (B-2-3) in the interval of the change in state from a certain point up to the parameters of the surrounding medium, we write

$$Ex \; = \; c_p (T - T_0) - T_0 \left(c_p \ln \frac{T}{T_0} - R \ln \frac{p}{p_0} \right) \tag{B-3-3}$$

Dividing the left- and the right-hand sides of eq. (B-3-3) by the value T_0, we obtain the equation in dimensionless values

$$\frac{Ex}{T_0} \; = \; c_p \, \frac{T}{T_0} - 1 \; - c_p \ln \frac{T}{T_0} + R \ln \frac{p}{p_0} \tag{B-3-4}$$

For satisfactorily calculating and constructing a diagram in dimensionless values, eq. (B-3-4) is written in the following form:

$$\ln \bar{p} \; = \; \overline{Ex} - \frac{c_p}{R} (\bar{T} - 1) + \frac{c_p}{R} \ln T, \tag{B-3-5}$$

where $\bar{T} = (T/T_0)$, $\bar{p} = (p/p_0)$, and $\overline{Ex} = (Ex/RT_0)$ correspond to the dimensionless temperature, pressure, and exergy of the material, respectively.

Using Eq. (B-3-5), values of the dimensionless pressure $\bar{p}$ are determined given different values of $\bar{T}$ and $\overline{Ex}$.

In Figure B-3-1 the scale along axis $\bar{p}$ is logarithmic, which is an additional convenience to the diagram since lines of equal values of the dimensionless

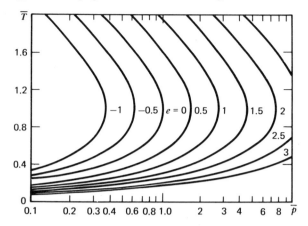

Figure B-3-1. Exergy diagram in dimensionless parameters according to Eqn. (B-3-5).

exergy are arranged uniformly in the field of the diagram. This permits obtaining the value of $\bar{Ex}$ from the diagram by dividing the intervals along the abscissa by any number of equal sections.

If an entropy diagram of state is available for the fluid used as a working body, it is convenient to construct a correction scale.

In Figure B-3-2 straight lines are constructed according to known methods [1, 5, 8] that consider the change in the temperature of the surrounding medium in the range $323-253\,^{\circ}$K.

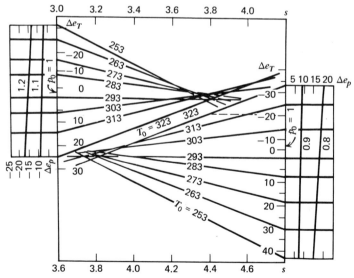

Figure B-3-2. Network correction scale for $e-i$ diagram for air (T, $^{\circ}$K; p, atm; kJ/kg $^{\circ}$K; e, in kJ/kg).

For determining the correction for different pressures of the surrounding medium, an auxiliary scale is constructed for the pressure correction where the value of the correction ΔEx_p is found from the following equation:

$$\Delta Ex_p = RT_0' \ln \frac{p_0'}{p_0} \qquad (B\text{-}3\text{-}6)$$

For more satisfactory calculations, the value of the grid scale for the correction is divided into sections: the upper and left parts are related to the value of entropy less than 4.2 kJ/kg°K^{-1} and a pressure of the surrounding medium of $p_0 > 1$ atm, and the right and lower parts are related to $s > 3.6$ kJ/kg° and $p_0 < 1$ atm.

Therefore, the total correction for a change in the parameter of the surrounding medium from T_0, p_0 up to T_0', p_0' is found by adding the two components:

$$\Delta Ex_\Sigma = \Delta Ex_t + \Delta Ex_p. \qquad (B\text{-}3\text{-}7)$$

The value of ΔEx_p is found with respect to the entropy value of the material at the point of the specified exergy and the line $T_0 = $ constant [1, 5, 8]. The value ΔEx_p is determined by the cutoff between the auxiliary lines p_0 and p_0 for a value of the temperature of the surrounding medium T_0.

As an example, we determine the correction from Figure B-3-2 for a change in the parameters of the surrounding medium from $T_0 = 293$ °K and $p = 1$ atm to $T_0 = 323$ °K and $p_0 = 0.8$ atm when $s = 4.4$ kJ/kg°K. The values are $\Delta Ex_T = -19$ kJ/kg and $\Delta Ex_p = 20.7$ kJ/kg. The total correction $\Delta Ex_\Sigma = -19 + 20.7 = +1.7$ kJ/kg.

NOMENCLATURE

In this appendix Ex is the exergy of the stream mass, i is the enthalpy of the mass, T is the temperature, p is the pressure, s is the entropy, c_p is the isobaric heat capacity, R is the gas constant, H is the altitude, ΔEx_p is the correction in determining exergy for the change in the pressure of the surround medium, and ΔEx_Σ is the total correction in determining exergy during change in the parameters of the surrounding medium. The index 0 relates to the pressure and temperature of the surrounding medium, the prime relates to the new value of the parameters of the surrounding medium, and the upper line denotes a dimensionless parameter.

SUMMARY

In this appendix the methods of determining the mass-flow exergy are given for changing parameters of the surrounding medium (temperature and pressure).

The dimensionless exergy diagram that is used for determination of the exergy values with any parameters T_0 and p_0 is presented in Figure B-3-1. A network of correction scales for the $Ex-h$ diagram for air that allows separate determination of corrections for changes of temperature and pressure of the surrounding medium is presented in Figure B-3-2.

REFERENCES

1. Brodyanskii, V. M., and Ishkin, I. P., *Kholod. Tekhn.* (1)(1962).

2. Keenan, I. H., *Mech. Eng.* (3)(1932).

3. Griun, G., and Kainer, E., Sb. "Voprosy termodinamicheskogo anliza" (collection, "Problems of Thermodynamic Analysis"). *Izd. "Mir"* (1965).

4. Rant, Z., Sb. "Voprosy termodinamicheskogo analiza," *Izd. "Mir"* (1965).

5. Meltser, L. Z., and Spinivasan, R. V. *Kholod. Tekhn.* (5)(1962).

6. Glaser, H., *Kaltetechmik* (11)(1965).

7. Korosholdt Christensen, B., *Kulde*, (1)(1965).

8. Kalinin, N. V., *Kholod. Tekhn.* (4)(1965).

9. Szargut, I., and Petela, R., *Egzergia (Warsaw)* (1965).

10. Fratscher, W., "Der Anhang zu M. P. Wakalowitsch, I. I. Nowikow," *Techn. Thermodyn.* (Leipzig) (1962).

11. Brodyanskii, V. M., *IFZh* (7)(1963).

12. Brodyanskii, V. M., and Ishkin, I. P., *IFZh* (10)(1963).

13. Marchal, R. *La Thermodynamique et la Theoreme de l'Energie Utilisable* (Paris)(1956).

Appendix B-4. Method of Thermodynamic Evaluation of Theoretical and Actual Cycles of Cooling Machines

L. H. MELTSER

Kholodilnaya Tekhnika i Tekhnologiya (6), Kiev (1968)

Thermodynamic analyses of quasistatic processes and cycles are used for analysis of actual cycles that are characterized by irreversible flow of all process, an increase in entropy, and the loss of exergy.

By designating the overall direction of the real process, classical thermodynamics will permit a quantitative determination of the increase of entropy and the losses associated with it only with throttling and the irreversible heat transfer for a finite temperature difference. In all other cases the deviation from the ideal cycle is determined empirically and is expressed by a coefficient whose value is very large. With the experimental efficiency it is possible to determine relatively simply the increase in entropy and to find the exergy losses. As a result, the overall exergy analysis of the real cycle can determine the total exergy loss, the rate of thermodynamic efficiency, and also the distribution of the total loss in percent relative to the specific processes that make up the cycle.

There has been a recent trend for wider use of exergy in engineering practice, and it is necessary to more clearly establish the feasibility of exergy analysis as a supplement to the conventional energy analysis.

OVERALL RATE OF EXERGY (THERMODYNAMIC) EFFICIENCY

The actual cooling cycle is characterized by a cooling coefficient ϵ and a rate of thermodynamic efficiency η. The values of ϵ for the same external heat source completely characterize the cycle since they all have the same reversible form. If

$$\frac{\epsilon_1}{\epsilon_{reversible}} > \frac{\epsilon_2}{\epsilon_{reversible}}$$

then for $\epsilon_1 > \epsilon_2$ the rate of thermodynamic efficiency is found to also correspond, that is, $\eta_1 > \eta_2$. For analyzing the reason that ϵ is lower than $\epsilon_{reversible}$ it is sufficient to look at the optimal conditions of reducing specific losses from a technico-economic point of view.

The analysis can be extended to a complex machine for which the specific percent of overall cooling production a_i is obtained for a defined external temperature condition.

For a complex plant consisting of many machines,

$$\eta = \epsilon \sum_{i=1}^{i} \frac{a_i}{i_{reversible}}$$

For the same temperature of the external heat source, the overall actual cooling coefficient ϵ of a plant completely determines the exergy quality.

Quite contrary to this is the comparison of cooling machines when working at different temperature conditions. In this case the comparison of the cooling coefficients is not a characteristic; therefore, it is necessary to compare the rate of exergy efficiency.

By using thermodynamic methods it is possible to compare any cooling machine working in different conditions. Therefore, it is even possible to consider machines for other purposes, for example, heat power, and in the general case any technological process connected to energy exchange. For example, Grassman [2] presented data on the rates of thermodynamic efficiency of home furnaces, the process of obtaining sodium carbonate, preparation of zinc, evaporation of salts, liquefaction, and even crushing of solid substances.

The author has composed an exergy table characterizing the thermodynamic efficiency of two related regions – cooling technology and power generation.

These general comparisons are useful only for developing the field and are inadequate in terms of exergy economy. There is an example we can use in the area of air conditioning where there is a large source of exergy loss in spite of the high value of the cooling coefficient.

From consideration of the exergy quality of plants for cooling technology and heat power, only one conclusion can be made. It is known that technico-economic calculations, based on determining the minimum operating expenditure and amortization of the costs, determine the level of admissable values of exergy efficiency. Relatively small variations in the value of the exergy efficiency for the widely used areas of energy and cooling technology apparently reflect the total cumulative mean static cost relationship between energy, water, metal, and operating expenses for maintenance.

REGION OF OPTIMAL VALUES OF η

In the past a specific type of cooling machine was used for a relatively narrow temperature region. Today the temperature boundaries used for each known type of cooling machine are significantly spread. In connection with this, it is useful to consider the mechanism of the variation in overall thermodynamic efficiency.

By presenting the overall rate of thermodynamic efficiency as the product of two coefficients $\eta_\Delta \eta'$, in which η_Δ characterizes the loss of exergy through external thermal irreversibility and η' through the internal transient processes, it is possible to show [4] that there exists an optimal temperature for which the given cooling generator has the best exergy characteristics η_{max} (see Table B-4-1).

Therefore, considering the overall rate of thermodynamic efficiency for a wide range of temperatures, it is possible to establish the regions of the optimum use of machines of different classes and types from the exergy point of view.

Considering the value ϵ in a wide range of temperature cannot give the optimal temperature since the cooling coefficient only increases when the lower temperature boundary is raised.

EXERGY-BALANCE LOSS

For developing recommendations on improving the action of the cycle, a balance of the exergy losses is made, in addition to the usual energy balance.

The loss of exergy in individual processes can be determined using one of the equivalent methods, entropy [3] or the exergy (continuous) [1]. When using the entropy method, the loss Π_i in a particular process is determined by the equation

$$\Pi_i = T_0 \, \Delta S_i,$$

where ΔS_i is the total change in entropy after completion of the process. The relative loss is calculated as the relationship of Π_i to the work capacity or exergy entering the cycle.

In the exergy method the idea of the exergy efficiency of the process is used:

Table B-4-1   Heat–Power and Cooling Technology

Plant	Effective exergy efficiency	Example
Heat and power		
Boiler plant of conventional electrical station	0.31–0.403	Power plants with 100–300-MW capacity
Binary mercury–water plants	0.455	Using improved methods; Upper temperature 565 °C
Gas plants	0.443	K-300-240 Plants and GT-10 on gas
Heavy diesel	0.38–0.40	
Gas Turbine	0.36	Turbine power to 100 MW
Gas turbine (peak)	0.31	
Cooling technology		
Single-stage vapor machine for + 10 °C and − 30 °C	0.25–0.30	Average data for machines working with different agents
Single-stage vapor machine working in region of maximum exergy efficiency (at − 10 °C)	0.4–0.45	
Two-stage vapor machine for − 30 °C and − 80 °C	0.45–0.1	
Two-stage vapor machine working in region of maximum exergy efficiency (at − 20 °C)	0.5	
Cascade vapor machine for − 80 °C to − 110 °C	0.3–0.1	
Philips type gas machine for − 70 °C and − 200 °C	0.3	Determined relative to reverse cycle maintaining constant temperature
Philips type gas machine for optimal temperature of − 150 °C	0.43	
Thermoelectric coolers for − 20 °C to − 23 °C	0.06–0.07	Quality of material $z = 3 \times 10^{-3}$ 1/deg.
Vortex coolers	0.01–0.02	

$$\eta = \frac{\text{Outlet exergy}}{\text{Total exergy}} = \frac{E_{\text{out}}}{E}$$

It is evident that $\Pi_i = (1 - \eta_i)E_i = T_0 \Delta S_i$.

The basic advantage of the exergy–entropy method of calculating the losses is the possibility of considering the disappearance of exergy as the result of irreversibility, whereas the energy analysis can consider only the unproductive expenditure of energy.

However, it is necessary to be careful when using the exergy-loss balance for determining "a narrow location" of the plant with respect to the value of greatest loss. Considering the exergy balance of an actual vapor-cooling machine, it is possible to conclude that significant exergy loss is related to the irreversible processes in the condenser. Improving this condenser does not give the required results since the loss strongly influences the work of the compressor, even if this compressor has an exergy efficiency near or equal to unity. It is necessary to always consider the divergence admitted when transferring from the ideal reversible cycle to the real irreversible cycle. The static exergy balance and the condition of relating the loss to a specific plant arrangement is an insufficient method for the exergy balance of the losses.

We examine the extent to which inefficient individual processes influence the overall cycle as a whole. If the particular efficiencies of the processes in the cycle are designated $\eta_1, \eta_2, \ldots, \eta_i$ and it is assumed that $\eta = f(\eta_1, \eta_2, \ldots, \eta_i)$, then from the equation

$$\eta = 1 - \sum \frac{(1 - \eta_i)E_i}{E}$$

in the first approximation, the following expression is obtained:

$$\frac{\partial \eta}{\partial \eta_i} = \frac{E_i}{E}$$

the efficiencies of the processes are considered independently of changes [5].

Therefore, it can be concluded that the cycle efficiency depends chiefly on the processes for which the relationship E_i/E has the greatest value. This principle can be used for evaluating theoretical cycles to show their sensitivity to the irreversibilities encountered in the conversion to a real system. From Figure B-4-1 the influence of the values E_i/E and η_i on the overall efficiency of the plant is seen.

A deficiency of methods based on composing and studying exergy-balance losses is the comparison with the reversible cycle. For study, a deformed irreversible process is already taken. The actual work in each irreversible process is compared with the reversible process for the same final conditions and not the final conditions in the ideal (100% isentropic efficiency cycle). Therefore, no matter what conditions were given for the ideal cycle, the thermodynamic analysis loses considerably.

Other types of analysis are also unsatisfactory in which an intermediate theoretical cycle is taken that consider only certain losses that are unavoidable for a given level of technology. It is appropriate to consider the actual cycle as a result of the transition: reversible cycle → theoretical cycle → actual cycle.

In this case the actual process should be considered in comparison with the corresponding processes of the reversible cycle, yet the possibility of forming

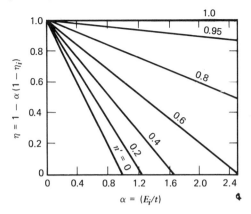

Figure B-4-1. Effect of individual values of exergy efficiency and specific exergy flow on overall machine exergy efficiency.

an analytical method to do this is very limited, since the process conditions are not the same.

GENERAL EXERGY ANALYSIS OF CYCLES

The analysis of a known and a particular new cycle requires developing methods of overall evaluation, considering the question of the reality of the theoretical indications after they are obtained. This question is not new. For example, it is well to know that the basic characteristic of the ideal Carnot cycle when actually realized is so degraded that a better actual performance is shown by other cycles with lower theoretical cycle performance but that do not change as much when transitioning to the actual conditions.

For analysis of the theoretical cycle, Martynovskii [6] used the criterion

$$x = \frac{\text{Theoretical work of expansion}}{\text{Theoretical work of compression}} = \frac{L_{T,e}}{L_{T,c}}$$

It was shown that the relationship of the actual cooling coefficient ϵ to the theoretical ϵ_T when $x \to 1$ approaches zero.

We consider the same problem from the exergy position. The overall cooling-machine cycle includes the process of regeneration of heat and work. Therefore, the exergy efficiency of any actual cycle can be represented in the following fashion:

$$\eta = 1 - \frac{L_c(1 - \eta_c) + L_e(1 - \eta_e) + E_h(1 - \eta_h) + E_b(1 - \eta_b) + E_0(1 - \eta_0)}{L_c - L_e}$$

where L_c and L_e are the actual work of the compressor and the expander (the efficiencies in the numerator have the values of exergy), respectively; E_h, E_b, and E_0 are the exergy of heat regeneration, cooling, and that lost to the surrounding medium, respectively; $(L_c - L_e)$ is the value entered into the exergy cycle; and L_c/E, L_e/E, E_h/E and so on are the relationship E_i/E for the individual processes.

We note that the values entered into the exergy cycle are equal in absolute values to the exergy of cooling in the reversible cycle.

From consideration of the expression for η it is possible to make the following conclusions:

1. The lower is the absolute value of E, the lower is the "exergy stability"[*] of the cycle. This indicates that the "stability" of the cycle is poorer when $x \to 1$.

2. Not only is the value x important, but also the absolute value of the work of compression and expansion, since for the same value of x and E the exergy efficiency decreases with an increase in absolute value in compression and expansion work. Therefore, for example, it is more economical to have a regenerative gas cycle in which the heat of regeneration reduces not only the absolute value of the compressor and expander work, but also the value of criterion x without changing the total exergy.

3. Lowering the value of x in the limit to zero (in the theoretical cycle) indicates a lowering of η of the actual cycle, but this lower value of efficiency has the greater stability for lower efficiencies of the individual processes.

4. For any cycle, it is possible to write the inequalities

$$\frac{L_c}{E} \geqslant 1; \frac{L_e}{E} \gtrless 1; \frac{E_h}{E} \gtrless 1; \frac{E_b}{E} \leqslant 1; \frac{E_0}{E} < 1$$

Hence it can be concluded that greater attention should be given to the improvement of the compression process and in cycles with regeneration, also to the process of regeneration.

Determination of the optimum variation of theretical cycles and the relationship between the mechanical and the thermal regeneration must be based on studies of sufficient values of efficiency for these processes and the interaction between these processes under actual conditions.

---

[*] "Stability" is defined in this case as the resistance to change when the theoretical cycle is converted to the actual cycle.

REFERENCES

1. Brodyanskii, V. M., "Termodinamicheskii Analiz Nizkotemperaturníkh Protsessov" (Thermodynamic Analysis of Low-temperature Processes), *IZD MEIm* (1966).

2. Grassman, P., *Eksergiya i Diagramma Potokov energii, Prigodnoi dlya Technicheskovo ispolzovaniya, "Voprosy Termodinamicheskovo Analiza"* (Exergy and Diagrams of Energy Streams Suitable for Technical Use, "Problems of Thermodynamic Analysis," Collection of translations under V. M. Brodyanskii, Ed.) "Mir," Moscow, 1965.

3. Hokhshtein, D. P., *Entropiinyi Metod Rasheta Energeticheskikh Poter* (Entropy Method of Calculating Energy Losses), Gosenergoizdat, Moscow, 1963.

4. Martinovskii, V. S. et al., "Energticheskaya Effektivnost Razlichnykh Generatorov Kholoda," (Energy Effectiveness of Different Cooling Generators.) *Kholod. Tekhn.* (2)(1961).

5. Martinovskii, V. S., and Meltser, L. Z., "Po Povodu Stati Ishkina i Brodyanskovo, *Kholodilnaya Tekhnika* (Concerning the report by Ishkin and Brodyanskii in *Kholodilnaya Tekhnika*), *Kholod. Tekhn.* (1)(1955).

6. Martinovskii, V. S., *Termodinamicheskie Kharakteristiki Tsiklov Teplovykh i Kholodilnhkh Mashin* (Thermodynamic Characteristics of Heat Cycles and Cooling Cycles of Machines), Gosenergoizdat, Moscow, 1952.

Appendix B-5. Thermodynamic Analysis of Reverse Lorentz Cycle

V. S. MARTYNOVSKII AND I. M. SHNAID

Kholodilnaya Tekhnika i Technologiya, (3), *Izdatelstvo "Tekhnika,"* Kiev (1966)

In the most technical installations using a reversible cycle the cooling medium is a stream of a fluid (water or air) with a finite heat capacity. (The term "heat capacity" is used in this report only relative to the total heat content of the body or stream. Its use here does not imply whatsoever the specific heat capacity of the substance.) In chemical technology, air conditioning, and other areas of technology there is widespread use of processes of cooling and heating a body or a fluid stream using the reverse cycle with various temperatures for supplying and rejecting the heat (the Lorentz cycle). With the aid of a thermodynamic analysis of this cycle it is possible to determine the minimum expenditure of energy and the thermodynamic features of many types of cooling machines and heat-pump plant.

Cooling in the conventional cycle is produced not only during cooling of the body, but in a stream of fluid with a variable temperature. Workers of cooling machinery at the Odessa Technological Institute of the Food Refrigeration Industry (OTIPKHP) conducted a thermodynamic analysis of the process of maintaining a steady-state regime for a cooling chamber [4]. The process of heat transfer in the insulation was considered along with the processes in the cooling machine that provided the conditions that would result in minimum irreversible loss in a closed thermodynamic system, including the surrounding medium, the cooling machine, insulation of the cold chamber, and the chamber itself. The solution of the variational problem formulated for this case shows that for a minimum irreversible loss it is necessary to use the Carnot cycle, withdrawing heat from the chamber, and the Lorentz cycle, removing heat from an intermediate temperature level of the insulation. The relationship between the cycle loads in the optimum case is completely determined by the temperature of the medium and the chamber. The total expenditure of energy in the cooling machine maintains a steady-state condition of a thermally insulated cold chamber when there is no internal heat evolution.

In practice, it is often difficult to bound the functions of one particular cooling plant. It is known that a significant part of the cooling load of a cold chamber is related to cooling during the temperature change.

We consider certain characteristics of the Lorentz cycle (Figure B-5-1). For

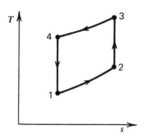

Figure B-5-1.   Reverse Lorentz cycle.

simplicity, we consider that the heat capacity of the process of supply and removal of heat is constant. In the calculations we assume the following designations:

$$c = \text{heat capacity of process } 1{-}2$$
$$c_h = \text{heat capacity of process } 3{-}4$$
$$q = \text{quantity of heat}$$
$$T = \text{absolute temperature}$$
$$S = \text{entropy}$$

$$a = \frac{c}{c_h}; \qquad t = \frac{T_1}{T_2}; \qquad t_h = \frac{T_4}{T_2}; \qquad t_0 = \frac{T_1}{T_4}$$

We consider the reverse cycle, which satisfied the second law of thermodynamics

$$\oint \frac{dq}{T} = 0 \tag{B-5-1}$$

if the temperature at the node points are connected by the relationship

$$t_h = t^a \tag{B-5-2}$$

Considering Eq. (B-5-2), the cooling coefficient ξ_a and the heating coefficient ϕ_a of the cycle can be presented as

$$\xi_a = \frac{c_{1,2}}{q_{3,4} - q_{1,2}} = \frac{1}{(1/at_0)\cdot[(t^{-a}-1)/(t^{-1}-1)]-1} * \qquad \text{(B-5-3)}$$

$$\phi_a = \frac{q_{3,4}}{q_{3,4} - q_{1,2}} \frac{1}{1 - at_0\cdot(t_h^{1/a}-1)/(t_h^{-1}-1)} \qquad \text{(B-5-4)}$$

Equations (B-5-3) and (B-5-4) are correct for any reversible Lorentz cycle independent of its technical capability to be practical and the properties of the working fluid. The only limit is included in the observed requirement for a constant heat capacity for processes 1–2 and 3–4. We analyze the expressions that have been obtained. When $a = 1$, the heat capacity of the process heat that is supplied and rejected is the same, and the cooling coefficient of the Lorentz cycle is

$$\xi_1 = \frac{1}{(1/t_0)-1} \frac{1}{(T_4/T_1)-1} \qquad \text{(B-5-5)}$$

Equation (B-5-5) agrees with the equations for the cooling coefficient of air-cooling machines, since the cycles of these machines are particular cases of the Lorentz cycle when $a = 1$. From the relationship in Eq. (B-5-5) it is seen that the cooling coefficient of the Lorentz cycle when $a = 1$ and the reversible Carnot cycle are essentially equal in the interval of temperatures $T_1 - T_4$. Consequently, for an infinitely large heat capacity of the surrounding medium, the use of the Carnot cycle for cooling a body does not lead to an overexpenditure of energy relative to the Lorentz cycle for $a = 1$.

Great interest is presented in the analysis of the characteristics of the Lorentz cycle when $T_2 = T_4$ and correspondingly, $t_0 = t$.

Such a cycle is conventional for thermoelectric, vapor-compression water, air cooling and other systems. When $t_0 = t$, the cooling coefficient of the Lorentz cycle is

$$\xi_a' = \frac{1}{(1/a)\cdot[(t^{-a}-1)/(1-t)]-1} \qquad \text{(B-5-6)}$$

With a decrease in a the cooling coefficient ξ_a' increases, reaching its maximum value when $a \to 0$:

$$\xi_1 = \lim_{a \to 0} \xi_a' = \frac{1}{[(\ln t^{-1})/(1-t)]-1} \qquad \text{(B-5-7)}$$

From Figure B-5-2 it is seen that the value of the energy expenditure during cooling is determined by the value of a to a great extent. (When constructing the graph, it was assumed that $t = 0.909$. Calculations show that the change

*It is natural to assume that in the Lorentz cooling cycle the initial temperatures are T_1, T_2, and T_4 and in the heat pump, T_1, T_3, and T_4; therefore, ξ_a is expressed by t_0 and t and ϕ_a, by t_0 and t_h.

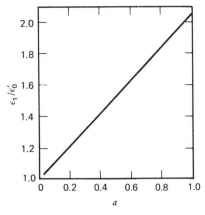

Figure B-5-2. Effect of value a on relationship ϵ_1/ϵ_a.

of t in the range 0.90–0.95 has practically no effect on the value of ξ_1/ξ_a'. Therefore, the graph in Figure B-5-2 is used for analyzing any standard cycle for water cooling or conditioning.) Such a characteristic effect of a on the energetic effectiveness of the cooling process is maintained in the thermoelectric water cooler and conditioner with a small temperature difference between the junction and the stream. Therefore, for raising the energy effectiveness of such a plant, it is necessary to increase the heat capacity c_h of the stream cooling the hot junction. The selection of a rational value of c_h must be based on the technico-economic calculations considering the connection between the value of c_h, the energy effectiveness of the plant, the cooling load at the hot junction, and so forth.

Because of the continuously lowering cost of electric power and the relatively high cost of water in modern small cooling plants with water condensers, the minimum operational expenditure corresponds to a small flow rate of cooling water and a large amount of heating of the water in the condensor ($10-14\,^\circ C$) [1]. Therefore, for a standard cycle of a cooling plant it follows to adopt the Lorentz cycle with a value of c_h determined by the flow rate of the water in the condenser.

It is interesting to compare the expenditure of energy in cooling a stream using the Carnot cycle, withdrawing heat at the low temperature, with the expenditure of energy when using the Lorentz cycle (Figure B-5-3). It is possible to be convinced of the fact that for the same cooling production there is less work expended in the Lorentz cycle, and the relationship of the expended work in these cycles is determined by the reverse relationship of the cooling coefficients:

$$\frac{\xi_c}{\xi_1} = \frac{t}{1-t}\left(\frac{\ln t^{-1}}{1-t} - 1\right) \tag{B-5-8}$$

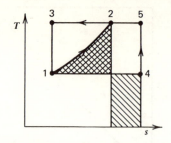

Figure B-5-3. Comparison of three-angle Lorentz cycle with Carnot cycle having same cooling-production removal of heat at lower temperature: $(1-2-3-1)$ Lorentz cycle; $(1-4-5-3-1)$ Carnot cycle.

where ξ_c is the cooling coefficient of the Carnot cycle withdrawing heat at low temperature.

It is obvious that

$$t \to 1 \lim \frac{\xi_c}{\xi_1} = \frac{1}{2}$$

$$t \to 0 \lim \frac{\xi_c}{\xi_1} = 0$$

Therefore, when cooling a body or a stream of fluid of constant heat capacity, the effectiveness of the Lorentz cycle surpasses the effectiveness of the Carnot cycle by no less than a factor of 2. This situation is well illustrated by the graph presented in Figure B-5-4. Equation (B-5-8) was used for its construction. From the graph it is seen that the use of the Lorentz cycle is expedient for any value of t, although the greatest preference from an energy standpoint is

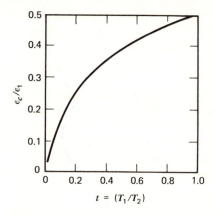

Figure B-5-4. Graph of function ϵ_c/ϵ_1 on t.

connectêd to cryogenic cooling (low values of t). For example, when cooling to a temperature of $280\,^\circ$K ($t \approx 0.93$) the expenditure of energy in the Lorentz cycle is 2.04 times less than in the Carnot cycle, and when cooling to $100\,^\circ$K ($t \approx 0.33$), the use of the Lorentz cycle decreases the expenditure of energy 3.03 times relative to the Carnot cycle.

For analyzing the Lorentz cycle in the simplest case when the heat capacity is constant, a series of sufficiently general principles is developed. In addition, the process of heating and cooling often requires assuming a constant heat capacity. Therefore, it is necessary in each specific case to determine the parameters and characteristics of the conventional cycle using basic thermodynamic concepts and the first and second principle of thermodynamics. From the cases considered above, it is possible to take the problem of determining the parameters of the standard cycle when there is intermittent change in heat capacity, as in the case of cooling a stream of fluid with a constant specific heat capacity and intermediate withdrawal at different temperatures. In such a case it is easier to divide the total standard cycle into several parts of a cycle with a constant heat capacity for the process of supply and rejection of the heat for which the functions presented in this report can be used.

REFERENCES

1. Ioffe, D. M., and Yakobson, V. B., *Malye kholodilnie mashiny i torgovoe kholodilnoe oborudovanie* (Small Cooling Machines and Commerical Cooling Equipment), Gostorgizdat, Moscow, 1961.

2. Martinovskii, V. S., *Kholodilnye mashiny,* (Cooling Machines), Pishchepromizdat, Moscow, 1950.

3. Martinvoskii, V. S., *Termodinamicheskie kharakteristiki tsiklov teplobykh i kholodilnykh mashin* (Thermodynamic Characteristics of Heat Cycles and Cooling Machines), Gosenergoizdat, Moscow, 1952.

4. Meltser, Y. Z., Shnaid, I. M., and Chainyakh, V. G., "Nekotorye zadachi nakhozhdeniya minimyma heobratimykh poter v kholodilnykh i energeticheskikh ustanovkakh (Certain Problems of Finding the Minimum Irreversible Loss in Cooling and Energy Plants"), Doklad na Vsesoyuznoi konferentsii na Termofisicheskie cvoictva veschestv, novye ckhemy i tsikly energeticheskikh yustanovok, Odessa, 1964.

5. Rosenfeld, L. M., and Tkachev, A. G., *Kholodilnye mashiny i apparaty* (Cooling Machines and Apparatus), Gostorgizdat, Moscow, 1960.

Appendix B-6. Combined Production of Heat and Refrigeration Using an Air-Cooling Machine

V. S. MARTYNOVSKII, L. Z. MELTSER, V. A. NAER, AND
L. F. BODNARENKO

Kholodilnaya Tekhnika (12), (December 1970)

Meat-processing plants expend significant quantities of fuel for obtaining a relatively low potential heat; therefore, it is advisable to evaluate different schemes of thermal transfer that can be used for the complex production of heating and cooling.

Different versions of conventional existing direct and indirect cycles are considered during the study of similar schemes. These cycles are composed of combinations of thermal transfer devices of the desirable type. Hence it is known in this case that the fuel economy is inevitably tied to supplementary expenditure by the equipment. Therefore, any engineering solution must be based on comparative analyses of the technico-economic effectiveness of the heat-transfer devices.

The principal thermodynamic aspect of the problem in the heat-transfer device is based on the classical works on the fundamentals of thermodynamics and a detailed analysis in many modern works [e.g., 1–3].

This important question was discussed not long ago in a report by Gorbatov et al [4]. However, there is a theoretical error in this report that cannot remain without consideration. The report discusses a new system in which the properties of an ideal gas are used for the working medium of an air-cooling system that is driven from a gas-turbine plant for "freely obtaining the cooling and electrical energy." It was further stated:

Actually, if the air cooling machine is made to work on the basis of heat consumption (analogous to the conventional thermal electric station), that is, to use a heat engine as the drive for the air turbocompressor while at the same time the heat of compression of the air in the air compressor and the heat removed from the heat of the gas engine are used for central heating, then the specific expenditure of energy for burning the fuel in developing a unit of cooling will be equal to zero. The air-cooling machine in this case becomes more economical relative to all cooling machines known at the present time.

Further discussions in the report led to conclusions that in the new scheme under consideration the development of heating and cooling is done "where the compressed air is obtained from an energy point of view free, that is, the compression of air in the final consideration does not consume any energy." It was concluded that "cooling is obtained as a supplementary product of the working plant without any additional energy consumption," and it was also stated that "the production of cooling changes from an energy requiring operation to an energy producing operation." As we see it, this terminology and the conclusions variously described the perpetual engine of the second kind.

The authors acknowledge that the results of their analysis are unusual; however, they write concerning their scheme, that "It permits making certain new scientific conclusions with reference to obtaining cooling on the basis of a synthesis of complex heat-cooling process." At the end of the report five general conclusions are given, each of which seems invalid.

After becoming acquainted with this work it is clearly disclosed that attempts by the authors to reason their conclusions lead to contradiction of the second law of thermodynamics. It is also necessary to emphasize that this mistake has remained in our scientific literature for a long time. Even in 1959 Masyukov [5] wrote, "We continue to claim by contrived means that mechanical work is obtained with the aid of a given heat-transfer device from the heat of the surrounding medium."

In the cited report the well-known general thermodynamic analyses were disregarded, including the accepted independent maximum possible effectiveness of heat-transfer devices (during reversible processes) based on the physical features of the working fluid. Neglect of this principle led the authors to a mistaken conclusion concerning the unusual properties of the air-cooling machine.

Finally, in the real cycle the properties of the working fluid are shown to have an effect on the actual characteristics of the heat-pump plant, but this cannot lead to a new conclusion regarding principles.

Any cooling machine, not only that for air, can, in principle, be used for obtaining cooling and heating on the basis of the well-known relationship

$$q = q_0 + l$$

where q is the heat produced, q_0 is the cooling production, and l is the work expended.

The coefficient of transformation $\phi = (q/l)$ depends on the boundary temperature of the cycle and the rate of its thermodynamic efficiency. The air-cooling machine is not excluded from this rule. The analysis of its energy characteristic leads to the conclusion that the air-cooling machine should not be used for obtaining refrigeration at temperatures around $-80\,°C$ and lower. In TCM-300 machines it is technically simple to simultaneously obtain a flow of hot air at a temperature of around $115\,°C$, which undoubtedly can be used for reducing the fuel expenditure and raising the overall economic use of the machine by 25–30%. However, it is impossible to omit the fact that all air-cooling cycles are characterized by a high value of the relative mean-temperature difference for the supply and exhaust heat that causes a low value for the coefficient for transforming the expended work into heat.

Finally, this shows that when the air-cooling machine is used, even in the most favorable region, it is necessary to consider using the most economical fuel.

In the report under consideration it was stated:

By using the heat from the air compression for central heating, the heat removed from the turbine gases and the heat from the water which cools the turbine combustion chamber, we find that all the heat of combustion in the fuel, considering certain losses, will be efficiently used. Consequently, the air which is compressed by the compressor in a given case will not contain energy obtained from fuel combustion.

This theory is based on a grave thermodynamic error.

First, instead of considering the whole cycle, which provides a continuous process of cooling production, we take individual elements. In the open thermodynamic process path, for example, the stored internal energy of the working fluid is used to obtain work (the adiabatic process of expansion). However, when using the complete combined processes that provide the continuous cycle, any attempt to not compensate for the "freely" obtained work and cooling is contradictory to the second law of thermodynamics.

Second, the different forms of energy are compared without considering their energetic value – their work capacity (or exergy).

The exergy of heat is determined by the well-known expression

$$E_Q = \int_1^2 T\,dS = T_0(S_2 - S_1).$$

The exergy of the working fluid expressing the maximum work that can be performed by the fluid during the transition from a state of nonequilibrium is expressed by the function

$$E_{P,T} = h - T_0 S$$

The change in this function determines the value of the maximum or minimum (depending on the process path) work during reversible transfer.

We consider the scheme proposed by the authors of the referenced report from a correct exergy position that leads to the following. The value of the exergy transfer of the working fluid during operation of the compressor is

$$E = h_2 - h_1$$

where h_2 and h_1 are the enthalpy of the air leaving and entering the compressor.

In the process of isobaric cooling the air after compression gives up only part of the exergy that it obtained, namely:

$$E_Q = h_2 - h_1 - T_0(S_1 - S_2),$$

where S_1 and S_3 are the entropy of the air before isobaric cooling and after it.

The compressed air transfers the remaining part of the obtained exergy

$$E_p = (h_3 - T_0 S_3) - (h_1 - T_0 S_1)$$

The balance is maintained since $E = (E_Q + E_p)$.

Therefore, the exergy analysis shows that the rationale in the authors' report concerning "free" compressed air is untrue.

When the initial source of energy is fuel, the principle for the distribution of the energy expenditure does not change. This elementary consideration from the exergy analysis was not considered in the authors report; therefore, errors in formulation were made, opposing the basic laws of thermodynamics.

In the exergy analysis presented in the preceding paragraphs the irreversible losses in the real process or the technico-economic side of the problem (cost of machinery and equipment, expenditure for maintenance, matching of heating and cooling requirements, etc.) were not considered.

It is completely obvious that the principle for combining fuel utilization was not taken by the authors in their report. This fact and the means of realizing the scheme lead to an unsatisfactory solution from both energy and technico-economic considerations.

For the drive for the air-turbocooling machine it is proposed to use a gas-turbine plant working with a higher exhaust gas temperature that can be used for producing steam or heating water. However, the use of the gas-turbine plant is connected to a larger loss than the direct cycle. The overall value of this loss is expressed by a lower thermodynamic efficiency that consists of around 25–27%, whereas in other thermal machines, such as the internal-combustion engine, the actual efficiency reaches 35–38%. Considering the increased temperature of the exhaust gas and the relatively small power, the efficiency of the gas-turbine plant in the proposed scheme probably does not exceed 20–25%. Besides this, it is uncertain whether the exhaust gases can provide the usual work necessary for use of a steam-generating boiler. In regard to this, with the appearance of a unitized energy system in the Soviet Union for which it is not economically favorable to have small electrical stations, the creation of many meat-processing plants with small self-sustaining power stations leads to an overexpenditure of fuel in general throughout the country.

The authors' scheme apparently did not give a value to the problem of increased efficiency for the direct cycle since they consider that the loss given up in the form of heat is equal to what would be used for the meat-processing plant. On this same matter, any irreversibility in the direct cycle for heat transfer leads to the fact that part of the initial expenditure of the work capacity of the fuel does not undergo efficient transformation and hence does not lead to economical fuel use.

The reversible cycle is also related to this problem where the air-cooling machine is proposed to be used for providing cooling to the relatively high-temperature storage chamber. It is known that in this case, because of the reduced relationship of the work of compression to the work of expansion, the value of the hydraulic and thermal resistance in the regenerator is increased and, together with a series of other reasons, the air-turbocooling machine has a lower value for the cooling coefficient.

The air-turbocooling machine can be efficiently used in regions where its energy characteristics approach or exceed the characteristics of the vapor-compression machine. As already pointed out, this temperature region is around $-70\,^{\circ}$C to $-80\,^{\circ}$C and lower. The use of these temperatures for freezing meat verifies the expediency of using the turbocooling machine since it significantly reduces the time of freezing, improves the quality of the product, and decreases wastage. This advantage is more substantial than fuel economy. Consequently, we have a reason to study the possible use of the air-cooling machine chiefly as a freezing device. During this, however, it should be kept in mind that for a meat-processing plant to use cold air, the frost must be removed from the chamber. Attempts to realize an industrial process during which frost is removed from a cooler by heating would approach a "isobaric" path. Successively passing through different sections of the meat-processing plant would require a higher-temperature cooler, which is impractical. The success connected with the requirements in obtaining the temperature and the main cooling production along different sections does not consider the loss in pressure and other construction and operating difficulties.

Therefore, a much more serious approach is required for evaluating the possible use of the air-turbocooling machine in a scheme for producing heating and cooling in the meat-processing plant.

For considering the question on using a heat-transfer device in the meat-processing plant, it is necessary to compare a series of schemes.

The scheme proposed by Gorbatov et al. [4] is not accompanied by even the simple calculations and construction considerations that are required for a qualified judgement on the engineering advantages.

The following versions of combined methods for obtaining heating and cooling are presented as having competitive capabilities in the general case:

1. A vapor-compression cooling machine used in the conventional scheme with an added compressor for a heat-pump stage to obtain hot water. In this case the fuel expended at the electrical power plant is utilized with maximum efficiency for developing electrical energy. The coefficient of transformation of the heat pump is high.

2. Using a vapor-compression cooling machine or an air-turbocooling machine operating with electrical energy from an electrical network and the heat from a thermal electrical station for the technical requirements of a meat plant.

3. Using an absorption cooling machine working from the heat of a thermal electrical station or from fuel.

4. Creation of a large meat-processing plant having a steam power plant with an excess back pressure (for thermal requirements) and a vapor cooling plant with a drive, for example, from a steam turbine.

Only a comparative and detailed complete technico-economic analysis of different versions of energy schemes and cycles combining the production of heating and cooling can resolve the most economical version in each specific case.

The assumptions and the theoretical basis for the report by Gorbatov et al. [4] seem invalid.

It is true that, in principle, it is possible to expend energy to obtain heat in a quantity exceeding its calorific capabilities and also to obtain work and cooling simultaneously. However, this was always known, as covered by problems in cooling technology and thermodynamics.

The miscalculation from the beginning to end was on the theoretical basis of the heat-pump principle.

For solution of the question about selecting the scheme for a combined thermal use system for a meat-processing plant, it is necessary to conduct special studies in which a thorough comparison of different versions must be conducted from a technico-economic position.

REFERENCES

1.   Kirillin, V. A., Cychev, V. V. and Sheindlin, A. E., *Tekhnicheskaya termodinamika* (Technical Thermodynamics), Energiya, Moscow, 1969.

2.   Lozhkin, A. N., *Transformatory tepla* (Heat Transformation), Mashgiz, Moscow, 1948.

3.   Martynovskii, V. S., *Teplovye nasosy* (Heat Pumps), Energoizdat, Moscow, 1955.

4.   Gorbatov, V. M., Gnoevoi, P. S., and Masyukov, V. N., "Otselesoobraznosti primenehiya no myasokombinatakh SSR vazdushnykh kholodilnykh mashin, rabotayushikh na baze

teplovogo potrebleniya" (Concerning the Question of Using an Air-Cooling Machine in Meat-Processing Plants in the Soviet Union Working on the Basis of Heat Consumption), Trudy Vsesoyuznogo nauchno-issledovatelskogo instutita myashoi promyshlennosti, Vyp. XXI. Moscow, 1968.

5.   Masyukov, V. N., "Kompressionnyi teplovoi transformator i ego termodinamicheskoe obosnovanie" (Compressor Heat Exchange and Its Thermodynamic Fundamentals), Vsesoyuzhyi nauchno-issledovatelskii institut mekhanizatsii selskogo khozyaistva, Izd. VASKHNIL, Moscow, 1959.

Appendix B-7. Thermodynamic Effectiveness of a Cooled Shield in Vacuum Low-Temperature Insulation

V. S. MARTYNOVSKII, V. T. CHEILYAKH, AND T. N. SHNAID

Isvestiya Akadamii Nauk SSSR, Energitika i Transport (2) (1971)

The conditions of maximum effectiveness have been considered for systems that include vacuum low-temperature insulation with an intermediate cooled shield coupled to a refrigeration machine. It is shown that such a system is very promising since it has an order of magnitude less loss of work capacity than does the conventional system. The optimal temperature and heat load of the shield are found.

INTRODUCTION

Joint consideration of the conditions of operating a refrigerator and low-temperature insulation permits calculation of the optimal energy relationship by technical solution. Although the final selection of the optimal version of the insulation must be derived considering economic factors, a purely thermo-dynamic analysis is required since it is an initial stage that cannot be bypassed for accurate calculation. This present work is devoted to a thermodynamic consideration of the combined action of a refrigerator and low-temperature vacuum insulation in which the heat is transferred by radiation.

The most universal method of decreasing the heat leak in cryogenic systems is by vacuum low-temperature insulation made up of a large number of reflective shields. For storage of low-temperature liquids (e.g., hydrogen and helium), systems are frequently used with vacuum insulation having the usual reflective shields (passive) and one or several active* shields, whose heat is generally

---

*By "active" we mean a shield that in a steady-state thermal regime has a total quantity of heat obtained from other elements of the insulation that differs from zero.

removed by the vapor of the stored liquid or by other means having a higher temperature [1–5]. Sometimes the active shield is connected to a separate cooler. Other low-temperature devices, particularly those where the cooling is developed at different temperature levels (e.g., in two-stage Stirling cooling machines), have recently been used in insulation construction with active shields.

Insulation with passive shields has been studied in rather extensive detail [1–3]. At the same time the effectiveness of introducing a cooled shield into the insulation has been studied with sufficient completeness only for the case of stored low-temperature liquids in a Dewar vessel where the shield is cooled by the vapor of the stored liquid [1–2].

In the present work a thermodynamic analysis is presented with a rather general case of vacuum insulation with active shields. Actually, the problem under consideration is specified for the continuous extension of the region of applied vacuum insulation with a cooled shield as the most effective. Earlier [6–8] it was shown that in conventional insulation with heat transfer by thermal conductance (in this case with vacuum-powder and vacuum-fiber insulation) the irreversible loss at low temperature can be substantially reduced by removing heat from the insulation at an intermediate temperature level. This present study is an extension of the work performed earlier and generalizes the previous results for the case of low-temperature vacuum insulation with radiation heat transfer.

BASIC RELATIONSHIPS

The degree of irreversibility in the insulation performance determines the value of exergy loss in the thermodynamic system, including the surrounding medium, the insulation itself, and the isolated cold body,

$$E = T_H \Delta S \qquad \qquad (B\text{-}7\text{-}1)$$

where T_H is the temperature of the environment and ΔS is the change in the entropy of the thermodynamic system.

If n active shields are arranged in the vacuum insulation with a temperature T_i,[†] the change in entropy is

$$\Delta S = \sum_{i=1}^{n} Q_i \left(\frac{1}{T_X} - \frac{1}{T_H}\right) + Q_x \left(\frac{1}{T_X} - \frac{1}{T_H}\right) \qquad (B\text{-}7\text{-}2)$$

where Q_i is the total quantity of heat given off by radiation of the ith shield. In the case of an active shield $Q_i \neq 0$ and if $Q_i > 0$, the heat from the shield is given up to a low-temperature liquid, a cooling machine, or some other source

[†] The numeration of the shields is taken sucessively from warm to cold.

of cooling. If $Q_i < 0$, the heat to the shield is given up by a thermal machine, a heat pump, and so forth. The values T_X and Q_X in Eq. (2) are the corresponding temperatures of the isolated cold medium and the heat leak to it from the insulation.

We assume, for simplicity, that the bounding surface and the active shield are parallel planes and that the surface emissivity C_n is the same for all surfaces of the active shield and the enclosure that exchange radiation. Then

$$Q_i = C_n F(T_{i-1}^4 - 2T_i^4 + T_{i+1}^4) \tag{B-7-3}$$

$$Q_X = C_n F(T_n^4 - T_X^4) \tag{B-7-4}$$

where F is the area of the radiating surface.

Some insulation constructions include k passive shields uniformly distributed in the space formed by the boundary of the surface and the active shield.* Then

$$C_n = \frac{C_k}{k/(n+1)+1} \tag{B-7-5}$$

where C_k is the emissivity for any surface pair exchanging radiation (it is assumed that for a given construction C_k = constant).

For a given value of T_H and T_X, the optimal temperature T_i of the active shield is that for which the minimum exergy loss is achieved, $E_X = E_{X\,min}$. In this case the temperature T_i must satisfy a system of n equations of the type

$$\frac{\delta E}{\delta T_i} = 0. \tag{B-7-6}$$

Using the Eqs. (B-7-1)–(B-7-4), we transform Eq. (B-7-6) to the form

$$4\left(\frac{1}{T_{i-1}} + \frac{1}{T_{i+1}}\right)T_i^5 - 6T_i^4 - (T_{i-1}^4 + T_{i+1}^4) = 0 \tag{B-7-7}$$

when $i = 1$ in Eq. (7) $T_{i-1} = T_H$, and when $i = n$, $T_{i+1} = T_X$.

Therefore, for determining the optimal temperature of the active shield, a system of n equations of the form in Eq. (B-7-7) is obtained that can be solved by approximation methods.

PROPERTIES OF OBTAINED SOLUTION

Analysis of Eq. (B-7-7) shows that

$$T_i^4 < \frac{T_{i-1}^4 + T_{i+1}^4}{2}$$

*It is natural that $k = (n+1)p$, where $p = 0, 1, 2, 3 \ldots$.

and in accordance with Eq. (B-7-3) $Q_i > 0$ at all times. Therefore, in the optimal insulation construction with a minimum exergy loss it is necessary to cool the active shield. When C_n = constant, its temperature and the heat load are defined by the boundary temperature and the value of n. It is characteristic that when C_n = constant, the optimal temperature T_i does not depend on the physical properties of the radiating surface. A schematic of the insulation construction with minimal exergy loss is shown in Figure B-7-1.

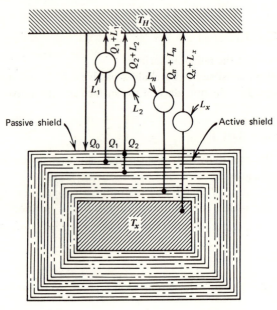

Figure B-7-1.   Principle of scheme for combined action of thermal insulation with cooling generator.

The conclusion obtained is not trivial, since earlier study of the functions in Eqs. (B-7-1)–(B-7-7) could verify only one conclusion; it is energetically unsuitable for a warm active screen with $T_i \geqslant T_H$. In the case when $T_i < T_H$ the energetic advantages of cooling the shield ($Q_i > 0$) against heating ($Q_i < 0$) is not so obvious.

The results obtained show that in the case of vacuum multilayer insulation with heat transfer by radiation, as in earlier studies of insulation construction with heat transfer by thermal conduction [6–8], the removal of heat at intermediate temperature levels with the aid of cooling machines decreases the exergy loss and raises the efficiency of the insulation. In the insulation with intermediate heat removal the heat flux from the surrounding medium increases, and the heat flux to the isolated cold body decreases relative to the insulation without intermediate heat-transfer removal.

We compare the energy effectiveness of the insulation with an active shield and the same system in which the shield is passive ($Q_i = 0$). For comparison of effectiveness it is possible to compare

$$\gamma = \frac{E}{E_0} \tag{B-7-8}$$

where E and E_0 are respectively the value of the exergy loss in the insulation with the active shield and the point in the same insulation where this shield is passive.

The value E_0 is determined from Eqs. (B-7-1) and (2), when $Q_i = 0$, and

$$Q_X = (Q_X)_0 = \frac{C_n}{n+1} F(T_H^4 - T_X^4) \tag{B-7-9}$$

Figure B-7-2 shows the dependence of γ with respect to T_i for the insulation with the single active shield ($n = 1$) for temperatures of $T_H = 300\,^\circ\text{K}$ and $T_X = 4.2\,^\circ\text{K}$, (the normal boiling temperature of helium). This curve has three sections bounded by points a and b. The part of the curve lying to the right of point b corresponds to $Q_i < 0$. At point b, $Q_i = 0$, and to the left $Q_i > 0$. Only to the left of point b can we have $\gamma < 1$.

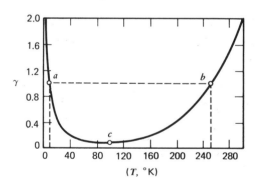

Figure B-7-2. Effect of active screen temperature on effectiveness of insulation construction.

It is interesting that cooling of the shield does not increase the effectiveness of the insulation for all temperatures T_i. It is obvious that the value $T_i < T_a$ is completely unsuitable since in this case the active cooled shield only degrades the effectiveness of the insulation ($\gamma > 1$). Point c corresponds to the minimal value $\gamma = \gamma_c$, and thus T_c is the optimal temperature for the cooled shield. From Figure B-7-2 it is seen that for helium temperatures $\gamma_c \approx 0.08$; that is, the use of even one active shield decreases the exergy loss more than 12 times. It should be emphasized that the value γ does not depend on the number k of passive shields arranged in the insulation. This means that even for perfect shielding of the heat flux, the use of one active shield permits decreasing the exergy loss by an order of magnitude in the considered case.

Figure B-7-3 presents the functions of γ_c and T_c with respect to $t = (T_X/T_H)$ when $T_H = 300\,^\circ\mathrm{K}$ for a system with one active shield. On the same curve is drawn the value of $\rho = [Q_X/(Q_X)_0]$. Figure B-7-3 shows that when reducing T_X the effectiveness of the system with the active shield increases simultaneously with a decrease in the value of T_c and p. For the optimal temperature of the active shield, the value ρ is identically determined by the values T_X and T_H. This shows that a sevenfold decrease in ρ, which sometimes is approached when constructing vessels for cryogenic liquid storage, can be justified either during the presence of a "donated" source of cooling for the active shield (e.g., the liquid vapor), or when there is required a sevenfold decrease in loss from stored products. If the effectiveness of the insulation is basically determined by the energy expenditure, it is necessary to proceed from the condition of minimal exergy loss and the sevenfold decrease in ρ is not justified.

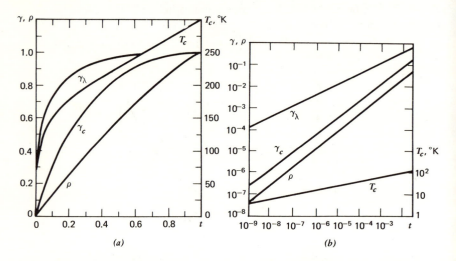

Figure B-7-3. Basic characteristics of insulation construction with one active shield.

Figure B-7-3 shows for comparison the dependence of $\gamma = \gamma_\lambda$ with respect to t for thermal conducting insulation with single heat removal and minimal exergy loss [6–8]. It is obvious that the use of intermediate heat removal in vacuum insulation gives the most effect.

Table B-7-1 presents the characteristics of vacuum insulation with different numbers n of optimal active shields when $T_X = 4.2\,^\circ\mathrm{K}$ and $T_H = 300\,^\circ\mathrm{K}$. Table B-7-1 shows that the increase in number of active shields leads to an increase in the effectiveness of the insulation (γ decreases). However, in most of the cases the technicoeconomic considerations make it expedient apparently

Table B-7-1   Vacuum Insulation with Different Numbers of Optimal Active Shields

n	T_c (°K)	$\gamma_c \cdot 10^2$	$\rho \cdot 10^3$
1	98	8.07	22.7
	180		
2	65	4.68	6.63
	220		
3	135		
	52	3.78	3.63

for $n \leqslant 2$. Attention is also drawn to the relatively high optimal temperatures of active shields for low-temperature cold bodies. This confirms the technical advantages of such a system with the use of a relatively simple refrigerator for cooling the shield.

The values of γ_c and T_c presented in the preceding paragraphs are related to the system with reversible cooling machines. Actually, the conditions of the thermodynamic efficiency rate of refrigerators decreases with lowering the temperature of the cooled bodies. Therefore, it should be expected that the actual value of γ_c will be higher than that considered; that is, the actual effectiveness of the system with the intermediate heat shield will be higher than the theoretical. Apparently, the optimal value of ρ for the actual condition must be somewhat less than that considered.

CONCLUSIONS

First, the removal of heat with the aid of a cooling generator at intermediate temperature levels from vacuum low-temperature insulation with radiation heat transfer is an effective means of decreasing the irreversible loss. This is true for any system with a shielded heat flow.

Second, for temperatures of the isolated medium down to helium temperatures (≈ 4°K) the advantages of an insulating system with an active shield is sufficiently accomplished, even with two active shields.

REFERENCES

1.   Scott, R. B., "Tekhnika nizkikh temperatur" (Low-Temperature Technology), *Izd. Inostr. Lit.* (1962).

2.   Kaganer, M. G., *Teplobaya izolyatsiya v teknike nizkikh temperatur (Thermal Insulation in Low-Temperature Technology)*, Mashinostroenie, (1966).

3.   Vance, R. W., and Duke, W. M., Eds. *Applied Cryogenic Engineering*, Wiley, New York, 1962.

4.   O'Banion, E., "Cryogenic Life Support System for Apollo," *Cryogen. Eng.*, 4 (4) (1969).

5.   Glaser, P. E., "Effective Thermal Insulation: Multi-Layer Systems," *Cryogen. Eng.* 4 (4) (1969).

6.   Martynovskii, V. S., and Shnaid, I. M., "Umenshenie neobratimykh poter v vysoko-temperaturnoi izolyatsii" (Decrease in the Irreversible Loss in High-Temperature Insulation), *Teplofiz. vysolykh temp.*, (5) (1964).

7.   Vikhorev, G. A., Meltser, L. Z., Cheilyakh, V. T., and Shnaid, I. M., "Energeticheskaya effektivnost metodov vnutrennego teplootvoda v izolyatsii kholodilnykh kamer" (Energetic Effectiveness Method for Internal Heat Removal in the Insulation of a Cooling Chamber), *Kholod. tekhn.* (1) (1966).

8.   Martynovskii, V. S., Meltser, L. Z., and Shnaid, I. M., "Teplovaya izolyatsiya obladayushchaya minimalnymi eksergeticheskimi poteryami" (Thermal Insulation Possessing the Minimum Exergy Loss), *Izd. Vuzov. Energ.* (10) (1966).

Appendix B-8. Combined Heating and Cooling Using an Air-Cooling Machine

V. M. GORBATOV, P. S. GNOEVOI, AND V. N. MASYUKOV

Kholodilnaya Tekhnika (8)(1971)

Martynovskii and his colleagues published a sharp criticism [1] of a report [2] concerning the use of an air-cooling machine working on the base heating requirements for a meat-processing plant.

Unfortunately, this criticism is based on the authors' lack of understanding of the physical sense of the proposed considerations and equations.

The author's letter to the editor selected a series of citations from our report, attempting to create the impression that we violated the second principle of thermodynamics and admitted perpetual motion of the second kind.

Actually, nothing in our report negated the second law of thermodynamics any more than for an air-cooling machine working on ideal gas.

At the same time, we considered only simplifying the discussions that the operation of the air-cooling machine was with an ideal gas, and we did not use the properties of the ideal gas as the basis for our conclusions and equations. We understand very well that according to the second law of thermodynamics the maximum possible thermodynamic effectiveness does not depend on the properties of the working body. In connection with this position of thermodynamics we never did betray neglect, as charged by the letter to the editor.

We now turn to the basic question concerning what we understand as "free" cooling electrical energy, air compression, and so forth.

Considering that theoretically all the heat of the burning fuel in the scheme of our plant is used in the electrical-power system network, we can maintain that the heat ΔQ quantitatively is not expended for obtaining electrical power and cooling. This is explained by the principle of energy conservation, and no one can deny this. However, the qualitative expenditure of heat ΔQ absolutely occurs, as explained by the second principle of thermodynamics. This proceeds in our plant scheme as a result of transition (transformation) of the heat ΔQ from the temperature level T_1 to the level T_2, where $T_1 > T_2$. Without such a compensating process it would be impossible to obtain electrical power and cooling, and we nowhere reject this.

In our report we write, "But in the system there is a compensating process expressing the transition of heat ΔQ from a temperature level of T_1 to a level T_2." For some reason the authors said nothing about this in their letter. They wrote nothing on our strict thermodynamic argument concerning the complete agreement of our conclusions and equations with the second law of thermodynamics.

Therefore, the compensating process or exergy expenditure (qualitative expenditure) of the heat ΔQ in our plant indisputably occurs, and this is clearly written in our report.

At the same time, the authors write in their report that "any attempts for noncompensation of the 'freely' obtained work and refrigeration is without sense because it is against the second law of thermodynamics," although this does not have the slightest relationship.

However, from this declaration by the authors in their report it follows irrefutably that Martynovskii and his colleagues understand the "free" receiving of work and cooling. They understand completely the decompensation of what has been obtained. By such an understanding of "freely" obtained work and cooling the authors attempt to add an absolute untruth.

Consequently, the chief charge in this matter is that we violated the second law of thermodynamics and propose the perpetual engine of the second kind in our plant, which is completely untrue.

The authors [2] do not wish to believe that all our conclusions and equations formulated on the basis of the first law of thermodynamics, are in complete agreement with the second law of thermodynamics.

The second law of thermodynamics prohibits the heat of the surrounding medium Q_0 to be changed into mechanical work with the aid of a direct thermodynamic cycle (postulated by V. Thomson).

Yes, actually such a change is completely impossible for any direct thermodynamic cycle because of the lack of a heat source with a lower sink temperature than the temperature of the surrounding environment. In the given case there would be no compensating progress. This violates the second law of thermodynamics since during a reversible process the process required to change any heat to work (as energy with a lower thermodynamic value, exergetically) is accompanied by complete compensation (the compensation process).

To clarify our system, (see Figure B-8-1) we establish the work for each cycle.: (1) the heat ΔQ_0 is changed into mechanical work $A \, \Delta L_{exp}$ of the

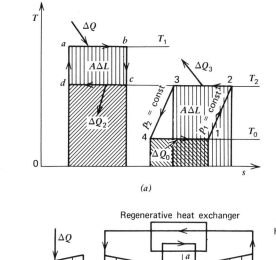

Figure B-8-1. Heat conversion cycle diagrams (*a*) and schematic (*b*) for the conventional and separated (along *aa*) systems.

expander; and (2) the heat of ΔQ is transferred from the source with temperature T_1 to a source with a temperature T_2. This is a completely compensated process.

We consider the work in each cycle for any inconsistency in the conventional exergy terms for the preceding effects.

First, during the process of change the heat Q_0 into work the exergy of the heat increases by the value Δe_1.

$$\Delta e_1 = A \, \Delta L_e - \Delta Q_0 \left(1 - \frac{T_m}{T_0}\right), \quad \text{but} \quad A \, \Delta L_d = \Delta Q_0$$

therefore,

$$\Delta e_1 = \Delta Q_0 - \Delta Q_0 \left(1 - \frac{T_m}{T_0}\right) = \Delta Q_0 \frac{T_m}{T_0}$$

but since

$$\frac{\Delta Q_0}{T_0} = \Delta s$$

then

$$\Delta e_1 = \Delta s\, T_m$$

Second, during the process of transferring the heat ΔQ from the source with temperature T_1 to the source with temperature T_2, the heat exergy decreases by a value Δe_2.

$$\Delta e_2 = \Delta Q \left(1 - \frac{T_m}{T_2}\right) - \Delta Q \left(1 - \frac{T_m}{T_1}\right)$$

or

$$\Delta e_2 = \Delta Q\, T_m \left(\frac{1}{T_1} - \frac{1}{T_2}\right) = -\Delta Q\, T_m \left(\frac{1}{T_2} - \frac{1}{T_1}\right)$$

Consequently,

$$\Delta e_2 = \frac{-T_m}{T_2}\left(\frac{T_1 - T_2}{T_1}\right)\Delta Q$$

but

$$\Delta Q \left(\frac{T_1 - T_2}{T_1}\right) = A\,\Delta L_c$$

Therefore,

$$\Delta e_2 = \frac{T_m}{T_2} A\,\Delta L_c$$

Considering that in our scheme the work of the gas turbine is expended only on driving the compressor, we obtain

$$\frac{A\,\Delta L_c}{T_2} = \Delta s$$

Therefore,

$$\Delta e_2 = -\Delta s\, T_m$$

Consequently, the overall change in the exergy of the system is expressed by

$$\Delta e_{\text{system}} = \Delta e_1 + \Delta e_2 = \Delta s\, T_m - \Delta s\, T_m = 0$$

hence

$$e_{\text{system}} = \text{identical,}$$

which completely corresponds to the second law of thermodynamics for reversible processes. Therefore, in our conclusions and equations there is nothing that contradicts the second law of thermodynamics, nor can there be.

We consider the work of the generally known gas-heat pump. If the heat of compression is completely removed after adiabatic compression of the gas, for each cycle of the heat-pump expander work will be performed because of the heat ΔQ_0. This follows from the energy-conservation principle. But the expander in the conventional heat pump is mounted on a single shaft with the compressor. Therefore, the heat Q_0 through the work of the expander and compressor passes into heat that is transferred to the heat sink at temperature T_2, where $T_2 > T_0$. Therefore, in the well-known heat pump the expander can work because of the heat ΔQ_0. The compensating process serves to change the heat of the engine into heat that also is transferred to the source at temperature T_2.

If the work of the expander in the conventional heat pump is not transferred to the compressor but generates current, the principles of thermodynamics are not disturbed. However, the electrical energy will be developed because of the heat from the surroundings ΔQ_0. On the basis of what has been shown, we cannot agree to the fact that our theoretical basis for the principles of the heat pump are untrue from beginning to end as stated by Martynovskii et al. [1].

It is necessary to note that there are many contradictions and inconsistencies between our report [2] and that of Martynovskii et al. [1]. For example, those authors [1] assume that in our report "instead of considering the whole cycle, providing a continuous process for the production of refrigeration, individual elements are taken." However, in our report two effects of working our plant are discussed: (1) on changing the heat ΔQ_0 into work of the expander (which creates the refrigeration effect) and (2) on transferring the heat ΔQ from the source at temperature T_1 to a source at temperature T_2. But these effects can appear only for each cycle of operation of the plant.

Furthermore, those authors [1] write that "the cooling machine is proposed for application to provide cooling of a relatively high temperature storage chamber," whereas our report [2] clearly states that in the experimental plant "it is expedient to use a turbocooler machine TCM-1-300." And it is well known that the machine provides cooling to $T = -80\,°C$ and lower.

Furthermore, those authors [1] state that "It is completely obvious that the principle of combined fuel use is developed by the authors of the report." But we did not pretend to author a generally-known principle. However, our principle simultaneously provides heat, cooling, and energy from the surrounding medium ΔQ_0. Nothing about this has been known, and thus it defends the authors' disclosure, as clearly stated in our report.

In the conclusion of their report [1] those authors contradict our conclusions. Actually, they write, "with the appearance of a unified state energy system in USSR under which it is economically unfavorable to have small electric power stations, the creation at many meat plants of small independent power plants leads to excess expenditure of fuel as a whole in the country." At the same time, the authors recommend for meat-processing a version of a cooling plant with the drive from an independent steam turbine.

Therefore, the criticisms written by Martynovskii and his colleagues [1] are completely invalid.

REFERENCES

1.   Martynovskii, V. S., Meltser, L. Z., Naer, V. A., and Bondarenko, L. F., "Combined Production of Heat and Refrigeration with the Aid of an Air Cooling Machine," *Kholod. Tekhn.* (12), (December 1970).

2.   Gorbatov, V. N., Gnoevoi, P. S., and Masyukov,V. N., "Concerning the Expedience of Using Air Cooling Machines Working on the Basis of Heat Input for Meat Plants in SSSR," *VNIIMP*, 21, Moscow, 1968.

Appendix B-9. Combined Processes in Cooling Technology and the Second Law of Thermodynamics

V. M. BRODYANSKII

Kholodilnaya Tekhnika (8) (August 1971) (Translated by John E. Ahern)

Improving the economy of different energy plants and systems, in particular those intended for simultaneous cooling and heating, is an importent task for the national economy. Therefore, all serious work of a theoretical and practical character leading in this direction must be given maximum support.

Together with this, it is necessary that each new proposal for creating economical schemes for cooling and heating be given careful substantiation from both thermodynamic and technicoeconomic considerations.

Taking this position, it is expedient to consider the questions raised in the discussions in the reports by Gorbatov et al. [1]. This consideration must necessarily be given sufficient detail so as to exclude any vagueness connected with different understandings of the same case. We begin the consideration with the plant scheme presented in that report [1]. This scheme (Figure B-9-1) has been taken without change from that report. For convenience of analysis, the gas turbine is separated from the remaining components by the dotted line 1−1.

It is not difficult to see that the proposed plant represents a conventional air-cooling machine (the part in the schematic to the right of line 1−1) with a drive from a gas-turbine plant (the part of the schematic to the left of line 1−1).

The gas-turbine plant will be capable of working either from a central heating plant "shaft," as proposed by those authors [1], or without it. Just so there is no doubt, the efficiency of the air-cooling plant is independent of the fact of whether the work of the expander is derived from an electrical generator (version 1) or directly from a turbocompressor (version 2). Nor is there any

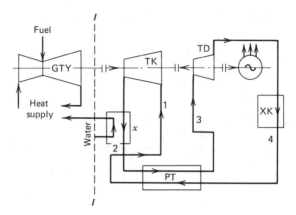

Figure B-9-1.   Schematic of a combined plant for refrigeration and heat supply according to Gorbatov et al. [1]: (GTY) gas-turbine plant; (TK) turbocompressor; (X) turbocompressor cooler; (PT) regenerative heat exchanger; (TD) turboexpander; (XK) cooling chamber.

doubt that the use of the warm water leaving the compressor cooler would be helpful.

All these technical solutions are repeatedly under consideration in practice, and they can be realized. Therefore, the efficiency of the proposed scheme is obvious and cannot be the subject of any discussion. The efficiency of the plant according to the scheme in Gorbatov et al. [1] was not contradicted in the report by Martynovskii et al [2].

The discussion related to another question, namely, whether the proposed scheme will allow the plant to be more effective than the conventional plant because of some principle advantage or whether it in fact has no advantage.

Gorbatov et al. [1] consider that such advantages exist by asserting that "the specific expenditure of energy of the burning fuel for generating a unit of cooling will be equal to zero." In emphasizing the advantages for the proposed scheme, the authors stated in a recent reply that "our principle simultaneously produces heat, cooling, and energy, in which the energy is produced because of the heat of the surrounding medium, ΔQ_0, which is not used and, therefore, only defends the authors' assertions that were clearly stated in our report." In addition, they consider that there is no violation of the thermodynamic principles in this conclusion also.

Martynovskii and his collegues [2] dispute the advantages of the proposed scheme and consider that the basis for these advantages is "wrong theoretically from beginning to end" and is contradictory to the second law of thermodynamics.

To make the consideration clearer, it is expedient to first exclude the engineering questions, which, although they play a substantial role, do not have a direct relationship to the principal positions that are the subject of this discussion.

The first question concerns the drive for the cooling plant. It is completely obvious that the principal feature of the proposed scheme does not depend on the type of drive if the drive energy is generated on the basis of heat consumption. For example, if an electric motor is substituted for the gas turbine, the electrical energy obtained from the thermal electrical stations is produced by heat supplied to the station, and in principle there is no change (by excluding this, the specific expenditure of fuel will become lower).

The second question relates to the temperature level of heat rejection from the cold object. It is obvious that in the temperature region below $-80\,^{\circ}\mathrm{C}$ to $-100\,^{\circ}\mathrm{C}$ the air-cooling machine is unsuitable since it will be poorer than the vapor-compression machine with respect to energy effectiveness. However, the question concerning the corresponding temperatures of the parameters for the generation and the necessary loss from cooling can be considered as the principal problem remaining.

We begin the analysis of the proposed scheme by considering the position in Gorbatov et al. [1] concerning the process in the air-cooling plant machinery – the compressor and the expander.

According to those authors [1], the process in the compressor of the gas-cooling plant is traced as follows:

In the air compressor the work obtained from the gas turbine is $A\,\Delta L$ and is equal to ΔQ_1 for adiabatic compression. This changes into a corresponding increment of heat content Δh and

$$\Delta Q_1 \;=\; A\,\Delta L \;=\; \Delta h$$

If this air compressed in the compressor is then cooled (at constant pressure) to its initial temperature, then a quantity of heat Δh or ΔQ_1 is removed from the air in the cooler.

Consequently, in this case the heat of compressing the air will be completely equal to the work expended by the gas turbine on compressing the air in the compressor.

Consequently, the compression of the air in the air compressor in this given case will not contain energy obtained during combustion of the fuel.

Here the authors state in a completely correct sense that the energy of compressing the air, as for any gas (if the deviation from ideal conditions are neglected), for the same temperature is the same as for incompressible air.

In Figure B-9-2a the schematic is presented on the energy balance for the process of gas compression in the compressor. From the energy balance it follows that the heat Q removed from the cooler is equal to the expended work L_K since the enthalpy I showing the energy of the gas stream at points 1 and 2

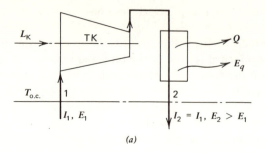

(a)

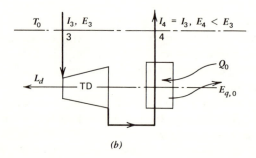

(b)

Figure B-9-2. Schematic of energy exchange in turbocompressor (*a*) and turboexpander (*b*):$T_{o.c.}$, temperature of surrounding medium; T_0, temperature of working body at exit from cooling chamber; L_K, work expended in driving turbocompressor TK; L_d, work delivered by turboexpander TD; E, exergy of gas stream; E_q, exergy of heat. Points 1, 2, 3, and 4 correspond to the schematic in Figure B-9-1.

is equal to

$$I_1 + L_K = Q + I_2, \qquad L_K = 0 \qquad\qquad \text{(B-9-1)}$$

The energy introduced into the compressor–cooler system in the form of work L is completely evolved into the form of heat Q.

From this correct premise, Gorbatov et al. [1] make a conclusion containing an incorrect condition, "In the considered case the compressed air is obtained free from the energy point of view, that is, in the final consideration the compression of air does not consume any energy."

This reasoning is an example of wrong conclusions that are based on the first principle of thermodynamics but do not consider the second law. Maintaining the requirements for the first law is a necessary but insufficient condition for practically any process. It follows to consider the conditions determined by the second law of thermodyanmics. Therefore, the "energy point of view" (correctly understood, obviously) must unavoidably consider not only the quantitative (first law) but the qualitative (second law) side of the energy change. From the quantitative point of view the energy during compression actually is not expended (it passes "through" the process), but the quality of the energy changes. The compression occurs with a reduction in the quality of energy – work L is entered, and heat Q is removed. This reduction in the quality of energy is the value that determined the expenditure by the compression.

To find a quantitative measure of the expenditure, we consider the exergy balance of the compression process (such a balance, as known, considers both the quantitative and the qualitative side of energy exchange).

The schematic in Figure B-9-2a is used:

$$L_K + E_1 = E_q + E_2 + D$$

or

$$L_K = E_2 - E_1 + E_q + D = \Delta E + E_q + D \tag{B-9-2}$$

From Eq. (B-9-2) it is seen that the work supplied L (or the exergy supplied) is expended by: (1) increase in exergy ΔE of the compressed air ($E_2 > E_1$, although $I_2 = I_1$), (2) exergy E_q of the heat loss, and (3) loss K through irreversibility (dissipation of energy).

Therefore, in the process of compression energy is not consumed but exergy (in the form of work) is, which contributes to the increase in exergy of both the compressed gas and the lost heat. The expenditure of exergy (L minus the exergy E_q of the heat if it is used) represents a necessary energy point of view for the process expenditure, which can be determined with reference to the fuel flow rate and economic value (in rubles). Therefore, there can be no talk of free compressed air here.

An analogous error is possible when considering the process in the turbo-expander, where it is maintained that "the work of the expander is obtained from the heat taken away from the air in the cold chamber" [1]. Here, as in the previous case from a correct consideration that the work of the expander L_E is equal to the quantity of heat Q_0 removed in the cooling chamber and determines this value, it is incorrectly concluded is made that the work is obtained from this heat (i.e., it is likewise "free") even more so since there is no refrigeration produced simultaneously. We examine this.

The exergy balance of the process in the turboexpander and the cooling chamber (Figure B-9-2b) has the form

$$E_3 = L_E + E_q + E_4 + D$$

or

$$\Delta E = E_3 - E_4 = E_q + L_E + D \tag{B-9-3}$$

Before analyzing Eq. (B-9-3) it is necessary to note the important case that when $T_0 < T_{o.c.}$, the value of the heat flow Q_0 and the exergy heat corresponding to it, E_q (the so-called reduced cooling production), have different signs [3–5]. Actually, since

$$\delta E_q = \delta Q \frac{T_0 - T_{o.c.}}{T_0} \text{ and } T_0 < T_{o.c.}$$

the heat flow and the exergy heat are directed onto the opposite sides. In other words, when the heat is taken from the cooling chamber, the exergy (the reduced refrigeration) is transferred to it from the working body. This exergy requires a corresponding expenditure and is specified in the final consideration of the cooling costs.

We return to Eq. (B-9-3), which shows that the work L_E of the expander and the cooling effect $E_{q,0}$ are provided by the expenditure of exergy (minus the loss D). This value $\Delta E = (E_3 - E_4)$ is created in the compressor and is transmitted directly to the regenerative heat exchanger at a level T_0.

Consequently, the two effects – the work of the expander and cooling – are the results of expending work that is obtained by the compressor from the gas turbine (or some other drive). It is impossible to obtain work "at the expense" of the heat Q_0. On the other hand, for creating a heat flow it is necessary to expend work, the value of which cannot be less than $E_{q,0}$, even in ideal cases.

The energy and the exergy balances of the system under consideration are shown qualitatively in Figure B-9-3 in the form of diagrams for the energy flow (Figure B-9-3a, Senki diagram) and the exergy flow (Figure B-9-3b, Grassman diagram).

Figure B-9-3a graphically shows the root of the error connected with considering the processes only from an energy balance. Actually, through the cross section $I–I$, energy obtained from fuel in the compressor drive does not pass, and the energy stream Q_0 removed in the cooling chamber is directed into the turboexpander. However, the latter transferred energy "itself" cannot

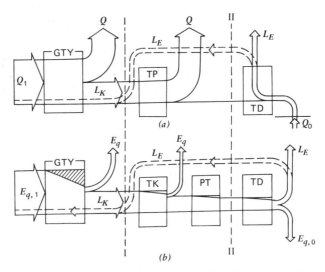

Figure B-9-3.   Energy (a) and exergy (b) diagrams of flow in scheme under consideration (designations are same as those for Figs. B-9-1 and B-9-2).

proceed; to do this it is necessary to expend energy. The distribution of these expenditures is represented graphically in the diagram by exergy streams (the cross-hatched triangles show the loss D of exergy in each part of the system). The work of the turboexpander and the cooling effect takes place because of the exergy flow $E_3 - E_4$ passing through cross section II–II, which in its own right is part of the exergy stream that starts from the fuel passing into the gas turbine.

Therefore, the "unusual conclusion," which states that the "cooling produced is converted from the energy required for production into energy produced" [1], is incorrect in principle and contradicts the second law of thermodynamics. The criticism of this and similar positions in the report by Martynovskii and his co-workers [2] is completely correct.

Certain works are necessary to comment on the reasons shown in which Gorbatov et al. [1] defend their own position. The equations for entropy balance in that work [1] and the exergy balance in the rebuttal [2] are correct in principle (if the inaccuracies in terminology and nomenclature are neglected). These equations show the efficiency of the proposed system and are actually not contradictory to the first or the second law of thermodynamics. But this is obvious without the equations. However, it does not follow at all from the equations that the proposed plant has a better effectiveness. Even more, it does not follow that "the production of refrigeration is changed from energy requirements into energy production" or as shown in the retort, "electrical energy will be produced because of the heat in the surrounding medium."

The latter assertion in the most clear form is shown by the authors' attempt to bypass the second law of thermodynamics and to obtain work from an absolute ineffective surrounding medium whose energy is not limited but whose exergy is equal to zero.

In a formal agreement with the position concerning the impossibility of obtaining work "because of the heat of the surrounding medium," Gorbatov et al. state that "the second principle of thermodynamics forbids the leakage of heat from the surrounding medium ΔQ_0, to be changed into mechanical work using a direct thermodynamic cycle (postulation by V. Thomson)."

V. Thomson did not verify any of the details; the reduced equation for the second law is in error and pertains apparently to the same author. The chief deficiency is related to the emphasis by the authors on the phrase concerning the direct thermodynamic cycle. From this it inevitably follows that to obtain work, only the average direct cycle is prohibited, but if any other cycle is taken or combination cycles are taken, work can be obtained. On this same matter, mention of the cycle in this equation is inadmissible, and thus to obtain work for a uniform surrounding medium is generally impossible for any cycle (or combinations of cycles). Any attempt to create such a cycle will soon reach failure.

In conclusion, it follows to mention certain words about priority. In their report Gorbatov et al. [1] assert that "the principle of simultaneously obtaining heat, refrigeration and energy in which the energy is produced because of the heat in the surrounding medium ΔQ_0 has never been known and therefore it only defends the authors assertions."

Unfortunately, this absolutely incorrect "principle" has been repeated in different modifications from the earlier authors [1], and different devices [6, 7] have been proposed.

Summing the reasons in the discussion, we can make the following conclusions:

1. The plant in the scheme due to Gorbatov et al. [1] is efficient in principle.

2. The criticism of the work of Gorbatov et al. [1], based on thermodynamic conclusions in the rebuttal [2], are correct.

3. The report by Gorbatov et al. [1] did not contain any new arguments that would refute the critical comments contained in the subsequent report [2].

REFERENCES

1. Gorbatov, V. M., Gnoevoi, P. F., and Masyukov, V. N., "O tselesoobraznosti primeneniya na myasolombanatakh SSSR vosduzhnykh kholodilnkh mashin, ratotayushchikh na base teplovogo potrebleniya" (Concerning the Expediency of Using on a Meat Combine in the Soviet Union an Air-Cooling Machine Working on the Basis of Heat Input), *VNIIMP,* **21,** Moscow (1968).

2. Martynovskii, V. S., Meltser, L. S., Naer, V. A., and Bondarenko, L. L., "Sovmentanaya vyrabotka tepla i kholoda pri pomoshchi vozdushnykh khololnykh mashin" (Combining the Work of Heating and Cooling with the Aid of an Air-Cooling Machine), *Kholod. Tekhn.* (12) (1970).

3. Brodyanskii, V. N., "Termodinamicheskii analiz nizkoteperaturanykh protssessov" (Thermodynamic Analysis of Low-Temperature Process), *NEI,* Moscow (1966).

4. Shargut, Ya., and Petela, P., *Energiya,* Moscow, (1968).

5. Sokolov, E. Ya., and Bodyanskii, V. N., "Energeticheskie osnovy transformatsii tepla i protssessov okhlazhdaniya" (Energy Basis of Transformation of Heat in Cooling Processes), *Energiya,* Moscow (1968).

6. Shelest, A. N., "Mashina atmosfernogo tepla" (Atmospheric Heat Machine) (1944).

7. Khalitrinov, V. P., "Transformator para s kipyatilnikom ammachnogo rastvora i absorberon vysokogo davleniya" (Transformation of Vapor from a Boiling Ammonia Solution with a High-Pressure Absorber) (1949).

Appendix B-10. Combining a Distillation Plant with a Gas Turbine

Z. P. BILDER AND E. I. TAUBMAN

Teploenergetika (2)(1973)

The use of a contact heat exchanger of the "gas–liquid" type in a multiple evaporation distillation plant removes the scale-forming process from the location of its most intensive formation – the head of the heater [1] or completely prevents scale formation on all heating surfaces of the plant.

The prevention of scale formation is accomplished by arranging the distillation process in such a way that the evaporating liquid does not contact the heating surface [2, 3]. This is achieved in particular by cooling the condenser of the multiple-staged evaporator by the distillate or another coolant that does not cause scale formation. Heat is transferred from the heat carrier to the air (or gas) and from it to the salt water in the contact heat exchanger.

The use of heat-exchange equipment for transferring heat from the distillate to the salt water through an intermediate gaseous heat carrier causes an additional exergy loss connected to the irreversible heat-exchange process in this equipment. However, raising the pressure and the corresponding temperature of the pre-heated salt water in the contact apparatus to a higher level than in the surface of the leading heater compensates for the loss and raises the thermodynamic effectiveness of the plant.

Figure B-10-1 presents a schematic of the power-distillation plant that operates with a GT-100-750-2 gas-turbine plant. In the plant having the scheme shown in Figure B-10-1a the exhaust from the gas-turbine plant heats the distillate in a recovery heat exchanger (12) and is then exhausted to the atmosphere. The distillate heats the air in another heat exchanger (13), cooling to a temperature of $45-50\,^{\circ}\text{C}$, and then is further cooled in the cooler (9) and

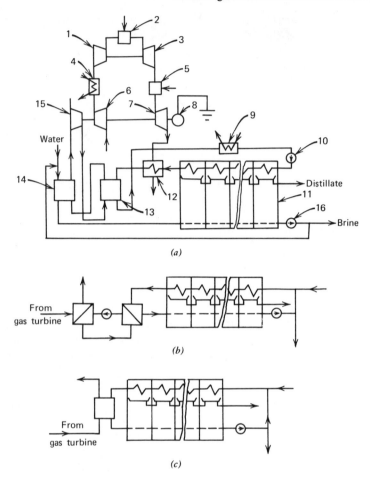

(a)

(b)

(c)

Figure B-10-1. Plant schematics: (1) high-pressure compressor; (2) high-pressure combustion chamber; (3) high-pressure turbine; (4) air cooler; (5) low-pressure combustion chamber; (6) low-pressure compressor; (7) low-pressure turbine; (8) generator; (9) distillate cooler; (10) pump; (11) staged evaporator; (12) heat exchanger; (13) air contact heater; (14) salt-water contact heater; (15) air blower; (16) pump.

pumped to the condensor of the multiple-stage evaporator (11) by the pump (10). The air from the heat exchanger (13) passes into the contact salt water heater (14), where it is cooled and then into the air blower (15), where it is pumped back to the surface or contact heat exchanger (13). Since the hydraulic resistance of the contact heat exchanger is not significant, the energy expenditure for driving the air blower is small.

Figure B-10-1 also shows a scheme for using a contact heat exchanger as a preheater for the multiple-stage evaporator (Figure B-10-1*b*). Figure B-10-1*c* shows a scheme for the multiple-stage evaporator with a surface preheater recovery unit and an intermediate heat carrier for the distillate in which the exhaust gas from the GT-100-750-2 gas turbine is used.

Figure B-10-2 shows the dependence of W/G_g (W-, production of freshwater distillation plant in m^3/hr; G_g, gas flow rate in tons/hr) with respect to t'' for different initial gas temperatures when heating the distillate in the recovery heat exchanger (12) by 25 °C. In addition, Figure B-10-2 shows the function

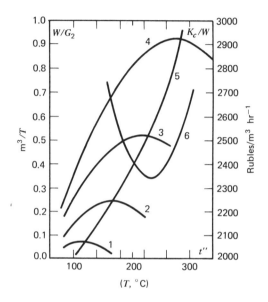

Figure B-10-2. Dependence of specific production of fresh water and specific capital cost of distillate plant as a function of the salt-water temperature entering evaporator. Initial gas temperature: (1) 500 °C; (2) 400 °C; (3) 300 °C; (4) 400 °C.

of the maximum plant production for different gas temperatures with respect to t'' for the scheme shown in Figure B-10-1 (curve 5). From the graph it is seen that with an increase in the temperature of the salt water in front of the evaporator t'', the production of the plant increases. However, for a constant flow rate and a fixed initial gas temperature, the maximum production of the plant is achieved at a specific temperature t. The temperature of preheating of the salt water, which determines the maximum production at a given gas temperature as shown by the calculations, does not depend on the temperature drop in the preheater. Figure B-10-2 also shows the specific capital costs K_c/W as a function of t'' (curve 6) of a distillation plant with a production of $W = 500$ m^3/hr (scheme in Figure B-10-1*a*) working on the exhaust gas from a GT-100-750-2 gas-turbine plant.

The capital cost of the distillation plant is determined in the following manner:

$$K_c = K_{mep} + K_{rhe} + K_{che} + K_{dc} + K_c$$

where K_{mep}, K_{rhe}, K_{che}, K_{dc}, and K_c, are the investment for the multiple evaporation plant, the recovery heat exchanger, the contact heat exchanger, the distillate cooler, and the air compressor (or blower), respectively. The value of K_{mep} is determined according to [4], K_{rhe} and K_{cd} according to [5] at the cost of the heating surface of 30 rubles/m^2, and K_{che} according to [6] at the cost of the regenerative heat exchanger of 60 rubles/m^3.

The specific capital costs, as seen from Figure B-10-2, has a minimum value at a temperature that corresponds with temperature t''. This temperature coincides with the maximum value of W/G_g.

For the comparison, calculations were performed on a water electric station composed of a GT-100-750-2 gas turbine and a water-distillation plant producing $W = 500$ m^3/hr of distillate and arranged in the third scheme in Figure B-10-1. During the calculations it was assumed that the temperature drop in the contact heat exchanger in Figure B-10-1a was 10% and in the preheater of Figure B-10-2b, 5%. The distillation plant used the exhaust gas from the GT-100-750-2 gas turbine at a temperature of $t'' = 240\,^\circ$C (Figure B-10-2). However, in a contact heat exchanger of the atmospheric type, the water can only be heated up to 80$\,^\circ$C, and thus in the scheme in Figure B-10-1b, $t'' = 80\,^\circ$C. In the multiple evaporative plant with the surface preheater, the temperature t'' was limited by the condition of a scale formation and was assumed equal to 100$\,^\circ$C. The results of the calculations are shown in Table B-10-1, from which it is seen that the specific capital costs of the plants under consideration are approximately equal. However, when it is considered that the plant in Figure B-10-1a operates without scaling of the heating surface, it follows that this one should be given preference. In addition, the specific expenditure of heat in this plant is lower than in the others. Here there is a reserve to increase the economy since the temperature of the exhaust gas from the recovery heat exchanger is equal to 200$\,^\circ$C. The use of this gas for preheating the salt water in a contact heat exchanger of the atmospheric type would give an additional quantity of fresh water from this plant. This gas can also be used to heat hot water for general supply by using a contact heat exchanger.

Raising the effectiveness of the electric station using a gas turbine with two compression stages can also be achieved by using the heat from the intermediate air cooler. Heating the distillate by air can be used in the plants arranged in the schemes shown in Figure B-10-1a, b; that is, instead of the recovery heat exchanger, the intermediate heat from air compression can be used.

The analysis of the schemes presented for a combined energy and distillation plant shows that the use of the "gas–air" type of contact heat exchanger is

Table B-10-1   Capital Costs

Plant characteristics	Schematic of plant in Figure B-10-1		
	a	b	c
Type of power plant	GT-100-750-2		
Production of distilled water, W, (m³/hr)	500		
Temperature of salt water before evaporator, t''' (°C)	200	100	80
Temperature drop in evaporator stage, Δt (°C)	25.80	16.35	11.70
Subcooling in stage Δt (°C)	2.1	3.9	3.1
Number of evaporator stages, n	6	3	3
Specific cost of distillation plant (rubles/m³ hr⁻¹	2405	2585	2400
Specific heat input for distillation, q (kJ/kg)	680	942	997

promising for heating salt water since it is possible to avoid scale formation on the surfaces of the distillation-plant heaters during high economic efficiency of the distillation process with a high rate of concentration in the solution.

REFERENCES

1. Bilder, Z. P., Kishnevskii, V. A., Lebedev, Yu. N., and Taubman, E. I., "Ispolzovanie contaktnykh teplo-obmennikov v kombinirovannykh energoopresnitelnykh ystanovkakh s gazovymi turbinami" (Use of contact Heat Exchangers in Combined Energy-Distillation Plants With Gas Turbines), *Teploenergetika* (5)(1971).

2. Maltsev, E. D., *Opresnenie solenykh vod.* (Distillation of Salt Water), Moscow, Atomizdat, 1965.

3. Dyablo, V. V., Zastavnyuk, V. K., Korheichev, A. I., and Kardasevich, O. P., "Termicheskaya opresnitelnaya ustanovka s gidrofobnym teplonocitelem" (A Thermal Distillation Plant with a Hydrophobic Heat Carrier), *Vodosnabzhenie i sanitarnaya tekhnika* (5)(1971).

4. Korneichev, A. I., and Dyablo, V. V., "Optimalnye parametry termoopresnitelnykh ustanovok" (Optimal Parameters of a Thermal Distillation Plant), *Vodosnabzhenie i sanitarnaya tekhnika* (8)(1970).

5. Dedusenko, Yu. M., Dedkov, G. V., *Optimizatsiya teplovykh skhem slozhnykh gazoturbinnykh ustanovok* (Optimization of the Thermal Scheme of a Complex Gas Turbine Plant), Kiev, Naukova Dumka, 1972.

6. Sosnin, Yu. P., *Gazovye kontaktnye vodonagrevateli* (Gas Contact Air Heaters), Moscow, Stroiizdat, (1967).

Index